T0320564

What Makes the Systems Engineer Successful? Various Surveys Suggest An Answer

What Makes the Systems Engineer Successful? Various Surveys Suggest An Answer

Howard Eisner

CRC Press
Taylor & Francis Group
Boca Raton London New York

CRC Press is an imprint of the
Taylor & Francis Group, an **informa** business

Library of Congress Cataloging-in-Publication Data
Names: Eisner, Howard, 1935- author.
Title: What makes the systems engineer successful? : various surveys
suggest an answer / Howard Eisner.
Description: First edition. | Boca Raton, FL : CRC Press/Taylor & Francis
Group, LLC, 2021. | Includes bibliographical references and index.
Identifiers: LCCN 2020037855 (print) | LCCN 2020037856 (ebook) |
ISBN 9780367545499 (hardback) | ISBN 9781003089650 (ebook)
Subjects: LCSH: Systems engineering—Vocational guidance. | Occupational surveys.
Classification: LCC TA157 .E37 2021 (print) | LCC TA157 (ebook) |
DDC 620/.0042023—dc23
LC record available at https://lccn.loc.gov/2020037855
LC ebook record available at https://lccn.loc.gov/2020037856

ISBN: 978-0-367-54549-9 (hbk)
ISBN: 978-1-003-08965-0 (ebk)

Typeset in Times
by codeMantra

This book is dedicated, first and foremost, to the author's wife, June Linowitz. June is an artist, and her creative process is much to behold. She also takes time out to support her husband's creative process. How? In many ways, but the primary one is just being there and paying attention.

Other family members provide support from time to time, and they deserve no less than an honorable mention. They are daughter and son and their spouses Susan, Oren, Joseph, and Tara, and my five grandchildren Jacob, Gabriel, Lee, Ben, and Zachary. I am interested in the reactions from all of the above to the notion of "success". Thoughts?

The book is also dedicated to the INCOSE Fellows who participated in the search and delineation of success attributes of the systems engineer. In particular, these Fellows and their comments are listed in Appendix A.

Contents

Preface

This book derives from the conjecture that the actual success factors for the systems engineer may be quite different from those suggested by INCOSE's certification approach. The actual factors, here called attributes, tend to be more personal and extensive, and they are believed to be measurable and teachable. So if the factors described here are accepted, or even partially so, that could well be the impetus for an initiative that deals with the factors and how they "fit" alongside of the INCOSE certification approach.

Basically, these new factors are developed by looking at clearly successful systems engineers and asking – what is it that is likely leading to their successes? And how is that answered? By looking at what they have achieved, as best we are able to do that.

Finally, the last point has to do with "success" and the decision to not struggle with the various possible definitions thereof. The author believes that success factors can actually be derived without being precise in terms of the definition of the word "success". So from this point forward, get ready to challenge the list of success factors. They were largely addressed by the author who has a Myers–Briggs Type Indicator with a very high "N".

Howard Eisner
Bethesda, Maryland

Author

Howard Eisner spent 30 years in industry and 24 years in academia. In the former, he was a working engineer, manager, executive (ORI, Inc. and the Atlantic Research Corporation), and president of two high-tech firms (Intercon Systems Corporation and the Atlantic Research Services Corporation). In academia, he was professor of engineering management and distinguished research professor in the engineering school of the George Washington University (GWU), Washington, DC, USA. At GWU, he taught courses in systems engineering, technical enterprises, project management, modulation and noise, and information theory.

He has written ten books that relate to engineering, systems, and management. He has also given lectures, tutorials, and presentations to professional societies such as INCOSE (International Council on Systems Engineering), government agencies (such as the Departments of Defense and Transportation, and NASA), and the Osher Lifelong Learning Institute (OLLI), American University.

In 1994, he was given the outstanding achievement award from the GWU Engineering Alumni.

Dr. Eisner is a life fellow of the IEEE (Institute of Electrical and Electronics), and a fellow of INCOSE and the New York Academy of Sciences. He is also a member of Tau Beta Pi, Eta Kappa Nu, Sigma Xi and Omega Rho, and various honor/research societies. He received a bachelor's degree (BEE) from the City College of New York, NY, USA (1957), an MS degree in electrical engineering from Columbia University, New York, NY, USA (1958), and a Doctor of Science (DSc) degree from the GWU (1966).

Since 2013, he has served as professor emeritus of engineering management and distinguished research professor at the GWU. As such, he has continued to explore advanced topics and write about engineering, systems, and management.

Other Books by the Author

There are nine entries under the title. These are:

- "Computer-Aided Systems Engineering"
- "Reengineering Yourself and Your Company"
- "Managing Complex Systems – Thinking Outside the Box"
- "Essentials of Project and Systems Engineering Management"
- "Systems Engineering – Building Successful Systems"
- "Topics in Systems"
- "Thinking – A Guide to Systems Engineering Problem-Solving"
- "Systems Architecting – Methods and Examples"
- "Systems Engineering – Fifty Lessons Learned"

Systems Engineering: No Room at the Top

1

This is a book about superior abilities and performance – not just better than average or passing a few tests that define a level of achievement. It is about seriously being the best as an engineer. We define "the best" by using real people as models, where the number one model is the ubiquitous da Vinci. There is essentially no person that will wish to argue that point, namely, that Leonardo is not the best. So what about just simply starting with him. Various comments will support the notion that this is a different time and a different place. These are "context" arguments that will not be neglected. So perhaps, to make the point we start with a definitive statement (which is not provable, but we accept in good faith):

Leonardo = Best systems engineer that ever lived

Rounding out the top are two other systems engineers, both of whom have some defects as super-systems engineers, namely, Newton and Einstein. What's the basis for his type of criticism for these two? Roughly speaking, it has to do with the limitations in scope of what they did, as we shall see later in this treatise. It certainly was not a limitation in brainpower. But, as we shall see, this is a complex story and is not all that easy to tell. But we have plenty of time and space.

So the overall perspective of this book is simply that we explore the set of (super) systems engineers and find out what they had in common as basic skills. This then becomes our answer: what is it that likely defines the success factors for this population of people? What are the success attributes in terms of conceiving of and building systems in the very best way. And how do we know what skills are employed by these super-systems engineers? That one is easy – they tell us, generally in the form of books and papers that they've written. That way we've got it in "black and white". And if they don't tell us directly, we may be fortunate enough to have a biographer. Much of the data is out there; it just needs to be collected and interpreted.

LEONARDO

Perhaps the most penetrating study of da Vinci is that provided by Gelb [1]. His investigation is complete and respectful, and winds up summarizing with the following seven points with respect to da Vinci's main attributes:

1. **An Insatiable Curiosity**. They all start with this; it provides the essential juices for discovery.
2. **Inclination to Test Knowledge with Experience**. If one's experience doesn't correlate with the theory, guess what?
3. **Enhance Senses, Especially Sight**. Makes elemental sense; seeing is believing.
4. **Accept Ambiguity and Uncertainty**. It's like being suspended in space, but he learned how to do it.
5. **Cultivation of Fitness and Grace**. Here's where the body comes into play directly.
6. **Develop Balance between Art and Science**. Some people call it having the right and left brains in balance, working to support each other.
7. **Systems Thinking**. This needs to be part of the overall approach to problem-solving, and it takes some explanation and practice.

Leonardo gives both himself and us this set of guidelines that can be followed. Each of us can look at this list of seven items and inquire –

• Can I look at this list and systematically improve each of the items?

In effect, we can then accept this list as his advice as how to become a better systems engineer.

We proceed here with three short anecdotes regarding da Vinci.

There came to be a time (1482 in particular) when Leonardo was applying for a job with the Regent of Milan. His letter was expansive, covering some 14 points and subjects such as his abilities in bridge-making, handling sieges, attacking fortresses, making cannons, catapults, and sculptures. I suppose it's not irrelevant to say; he got the job. And by the way, he also added an item that addressed the matter of making peace, just in case the Regent got the impression that he was some type of warmonger.

From 1495 to 1498, he was working on "The Last Supper", capturing the scene in which Christ declares that "one of you will betray me". This is a singular moment from a singular artist. The anecdotal part of this incredible piece of art is the fact (prediction) that he knew he would be betrayed and also perhaps knew who would betray him.

If we look carefully at the seven da Vinci principles, as cited above, we see his broad skills and thinking, summed up, if you will, by "systems" thinking. It is not a far leap of faith to see that da Vinci was actually the originator of this type of thinking, important today to the systems engineer.

Finally, yet another "saying" is attributable to da Vinci: one that will find special resonance in this treatise. This saying may be loosely translated as "every obstacle that I find in front of me will yield to stern resolve".

NEWTON

Born in 1643 in the UK, he became the time's preeminent physicist, mathematician, astronomer, natural philosopher, and theologian. He is possibly best known for his equations of motion as well as his formula for the gravitational force. He also contributed in a fundamental way to the fields of calculus and optics. He was much more than a theoretician and academic, having invented the reflecting and Newtonian telescopes. Here are a few well-known sayings that are attributed in the literature to Newton [2,3]:

a. "If I have seen further than others, it is by standing on the shoulders of giants".
b. "We build too many walls and not enough bridges".
c. "If I have done the public any service, it is due to my patient thought".
d. "To every action there is always an opposed equal reaction".
e. "I now demonstrate the frame of the system of the world". ("Principia Mathematica").

We follow this with three anecdotal stories about Newton. The first is simply refuting the "apple" story. Many think that Newton came upon his theories of gravitation by observing an apple falling from a tree. According to a background treatise [2], this apple story is not true.

A second anecdotal area is that Newton held the position of Master of the Royal Mint. He worked on that position with skill and fervor, leading to the conviction of many as counterfeiters. The penalty he argued for was being hanged, which he obtained in many cases. One guesses that he received no small measure of satisfaction from getting a good count on the hangings.

Newton was a vindictive as well as unrelenting man, which he demonstrated in his Mint position as well as an intense set of arguments with the likes of Hooke and Leibniz. In other words, he seemed to love a fight. According to a biography [2], he was not inclined to back up or give up when a controversy was at hand. His tendency toward perseverance carries over to our benefit – namely, we still use Newton's genius today to solve certain gravitational problems. Thank you, Sir Isaac! And keep in mind that there is yet another piece of work from Sir Isaac that transcends just about any and all treatises written to that day. And that is Newton's *Principia* [4].

EINSTEIN

Along with da Vinci and Newton at the top of the profession is Albert Einstein [5], who was actually a rather playful individual, with a deep regard for humor and humility. He formulated the breakthrough special and general theories of relativity, visualizing a space-time continuum. He set forth the well-known $e = mc^2$ relating mass and energy. He believed that the use of one's imagination was more important and useful than pure knowledge and facts. He set forth a two-stage combinatorial play process that helped him in his problem-solving. He also used gedanken experiments to work his way through difficult sequences of logic and complex behaviors.

Einstein had a multi-faceted life, starting if you will as a patent reviewer in Berne, Switzerland. He was modest in his early day achievements and did not appear to be the shooting star of physics that he eventually became. Perhaps a turning point was the solar eclipse of 1919 when his relativity theory was, in part, demonstrated. This had to do with the bending of starlight by the sun, by some 1.4 arcseconds (he had predicted 1.7 arcseconds). Apparently, the world was also watching, and they had found their physics genius; he had arrived with that one measurement. He had persevered.

According to this author, he deserves the right to be acclaimed as a super-systems engineer, even though he did little by way of running a huge organization. But his influences were manifold, which we all took notice of, year by year. He was even offered a serious leadership position in Israel (which he turned down), operating as a world-wide purveyor of wisdom and good will.

Various approaches and thoughts of significance in relation to this subject and Einstein can be paraphrased as follows:

- Prepare, from time to time, to break all the rules.
- I have often been seen as one who has contempt for authority.
- All of what we call science is simply the application of good sense.
- I prefer visualization to trying to think in thoughts.
- My curiosity is like a little plant that is in need of freedom.

Following the approach with Messrs. Da Vinci and Newton, here are three very short stories about Einstein. That completes this introduction regarding our very best systems engineers – da Vinci, Newton, and Einstein. We move on down to more modern people and times in the following chapter.

One of the more curious stories about Einstein by way of an anecdote is his tendency to not wear socks. After a while, that just became his way of being, so he didn't notice that it was not the norm. Other people did, of course.

A second area of some note with respect to Einstein is the fact that he seriously engaged in "combinatorial play". Doing so, he found that his creativity was enhanced. Of course, looking at all (or most) of the combinations is also one way was synthesizing new systems and solutions to problems.

Finally, in a letter to French mathematician Hadamard, Einstein apparently said that words, or language, do not seem to play an important role in his thinking. Rather, pointing to the above, he believed that the real answers lie in combinatorial play and visualization. We could certainly do a lot worse than copy Einstein in terms of what he did and how he thought and engaged in problem-solving.

REFERENCES

1. Gelb, M., "How to Think Like Leonardo da Vinci: Seven Steps to Genius Every Day", Dell Trade, 1998.
2. Kolme, Korkeude, Uy., "The Entire Life Story of Sir Isaac Newton", The History Hour Press; see www.thehistoryhour.com.
3. Kolme, Korkeude, Uy., "Sir Isaac Newton – One of the Greatest Minds of All-Time", The History Hour Press; see www.thehistoryhour.com.
4. Isaac Newton with Cohen, B. "The Principia", The History Hour Press; see www.thehistoryhour.com.
5. Dorminey, B., "Lesson from Einstein: Genius Needs Perseverance", *Forbes*, 23 October, 2016.

Selected Best Systems Engineers

2

We take a brief look, in this chapter, at a variety of systems engineers that have been, and are, at the top of the profession. If we examine their lives, we see what it takes to be successful systems engineers. These are thumbnail sketches that suggest further exploration and examination.

SYSTEMS ENGINEERS RUNNING LARGE GOVERNMENT ORGANIZATIONS

- **Admiral Rickover**. Hyman Rickover was the well-established Navy director of the Nuclear Submarine forces [1]. This capability represented one-third of this country's defensive capability under the MAD (mutual assured destruction) policy and concept.

  As an elder statesman, two comments attributed to him have been: I am not proud of the part I played but the war which I did because it was necessary for the safety of our country, and if you're going to sin, sin against God, not the bureaucracy. God will forgive you, but the bureaucracy won't.

- **Robert Oppenheimer**. Oppenheimer [2] was the civilian manager of the "Manhattan Project" whose charter was building the first atomic bomb. This critical undertaking required massive leadership and intelligence skills which Oppenheimer possessed, and thus, he was a major contributor to the war's success against two very dangerous enemies (i.e., Germany and Japan in WWII).

Dr. Oppenheimer, despite his position and accomplishment, did not support the creation of a hydrogen bomb. As a result, many hawks attacked him. Nonetheless, he received the Enrico Fermi award from President Kennedy, and due to the assassination, from President Johnson. He became connected to Hinduism and was known for its quote "Now I am become Death, the destroyer of worlds".

- **Werner von Braun**. Werner von Braun headed the German team in pursuit of the atomic bomb. At the conclusion of the war, he settled in the United States and contributed significantly to the rocketry program in the States. He was well-known for his contributions at the Missile Command in Huntsville as well as other NASA centers such as Goddard Space Flight Center and Headquarters.

Dr. von Braun played an important role in Germany and, after the war, was able to be committed the United States. That commitment took several forms such as the Explorer 1 satellite, the building of the Space V super heavy lift vehicle and director of the Marshall Space Flight Center. He was accepted into the National Academy of Engineering and received the National Medal of Science.

SYSTEMS ENGINEERS WITH STRONG ACADEMIC BACKGROUNDS

- **Andrew Sage**. Sage brought systems engineering from the classroom to the country at large. He was the Editor-in-Chief of INCOSE's *Systems Engineering* journal which was in itself a major contribution. His interests ranged far and wide, and included software engineering and systems of systems engineering. He served as the first Dean of systems engineering at George Mason University.

Dr. Sage was recognized as a Fellow of INCOSE and a Life Fellow of the IEEE. He was elected to the National Academy of Engineering and received honorary doctorates from the Universities at Waterloo and Dalhousie. He was awarded the INCOSE Pioneer Award, the IEEE Simon Ramo Medal, and the IEEE Donald Fink prize paper award.

Andy Sage put forth a lot of effort helping other practitioners of systems engineering and moved into software engineering when he thought it was appropriate.

- **Wayne Wymore**. Dr. Wymore made seminal contributions to systems engineering by stressing mathematics and modeling. He was the initiator of Model-Based Systems Engineering (MBSE). He served as the first Chairman of Systems and Industrial Engineering at the University of Arizona. He was recognized as a Fellow by INCOSE.

  Dr. Wymore wrote four important texts in systems engineering, oriented toward modeling and design.

- **Ben Blanchard and Wolt Fabrycky**. The powerhouse of these two colleagues at Virginia Polytech was recognized everywhere. Their book on systems engineering and analysis was accepted as a leading-edge contribution to the overall field. They brought their University into the modern times of systems engineering and systems analysis. Both are Fellows of INCOSE (Pioneer Award).

  Dr. Blanchard, in addition to his significant contributions in academia, worked in industry for Boeing, Sanders, Bendix, and General Dynamics. He has since passed.

  Dr. Fabrycky has also received Fellow status from the Institute of Industrial Engineers, the American Society for Engineering Education, and the American Society for the Advancement of Science. He has received several awards, including the Holtzman Educator Award, the Lohmann Medal, and the Wellington Award.

- **Henry Petroski**. He was a Professor at Duke, and a prolific writer (12 books). He was recognized as expert in his field of failure analysis and also as creative leading-edge teacher, clearly a leader in his profession (civil engineering and history), and a receiver of several honorary doctorates.

  Dr. Petroski received many awards including the Cosmos Club Foundation award, the Product Safety Design Award from ASME, the Washington Award, and the Lecture Award from ASCE, and was recognized as a Distinguished Member of the ASCE.

SYSTEMS ENGINEERS FROM MULTIPLE SECTORS

- **Eberhardt Rechtin**. Eb Rechtin was a three-sector man: President of the Aerospace Corporation, Professor at USC, and Director of DARPA (Defense Advanced Research Projects Agency) (industry, academia, government). In addition, he wrote the seminal book on systems architecting, and applied some of his concepts to organizational behavior.

Dr. Rechtin received several awards including the Navy Distinguished Public Service award, the AIAA Goddard Astronautics award, the Japan C&C Prize, and the AFCEA Gold Medal for Engineering. He was a member of the National Academy of Engineering and was given the title "Father of the Deep Space Network".

* **Norman Augustine**. Augustine, an aeronautical engineer and also a three-sector contributor, occupied leading positions in government (Acting Secretary of the Army), industry (president of Lockheed Martin) and academia professor (Princeton). He served as key advisor to government on matters of defense, business, transportation and technology. He wrote two books, one commenting on his "laws" and the other on his "travels".

  Norman Augustine has received numerous awards and accolades, to include the Eta Kappa Nu Eminent Fellow, the Vannevar Bush Award, the ASME Medal, the von Kaman Wings Award, the Air Force Public Service Award, and the NAS Award in Aeronautical Engineering. He served as Chairman of the National Academy of Engineering and the Defense Science Board. He received several honorary doctorates from a variety of universities. He took lead positions with the Red Cross and the Boy Scouts. He is one of this country's most capable engineers and managers.

* **Jacobs and Viterbi**. Key technologists and businessmen. For Jacobs, he got his doctorate at MIT and became the CEO of Qualcomm, a high-tech communications company. For Viterbi, after he and Jacobs joined early forces at Linkabit, he created multiple access communications technology. Both were very strong technologists and held positions as academics.

  Dr. Jacobs is a Fellow of the IEEE and a member of the National Academy of Engineering. His awards included the Inventing America's Future award, the Cornell Entrepreneur of the Year award, the IEEE Graham Bell Award, and the Woodrow Wilson Award. He was inducted into the National Inventor's Hall of Fame

  Dr. Viterbi had the school of engineering at USC named after him. He has received numerous awards including: the John Fritz Medal, the Benjamin Franklin Medal, the IEEE Medal of Honor, the National Medal of Science, the IEEE James Clerk Maxwell Medal, the Claude Shannon Award, the Marconi Prize, and the IEEE Alexander Graham Bell Medal.

SYSTEMS ENGINEERS FROM STRONG BUSINESS BACKGROUNDS

- **Tom Watson, Jr.**, Watson led the charge at IBM, establishing a "thinking" mindset and culture. The latter were based largely upon three notions: (1) respect for the individual, (2) that the company provided the best possible service for the customer, and (3) constant pursuit of excellence and superior performance. Of course, it did not hurt that during Watson's tenure the company introduced the highly successful 360 series of computers and emphasized the function known as marketing.

  Tom Watson was the president of IBM from 1952 through 1971. He also served as ambassador to the Soviet Union. He was awarded the Presidential Medal of Freedom by Lyndon Johnson. Time magazine declared him to be one of the 100 most influential people of the 20th century.

- **Hewlett and Packard**. Here's another one of those garage start-up stories. Friends and serious engineers thought they could do corporate engineering, and they did so. A phrase from one of their annual reports – we are well positioned to create value for our customers and stockholders, harnessing the power of in the formation, the way people live, and how businesses operate. HP documented their philosophy in a treatise called the HP Way.

  Hewlett and Packard won and created many awards, and led the charge in such areas as a commitment to innovation, listening to customer, management by walking around (MBWA), flexible organizational structure, and trust in people.

- **Andy Grove**. Andy Grove emigrated from Hungary and obtained BS and doctorate degrees in the States at CCNY and UC Berkeley, respectively. He worked with Robert Noyce and Gordon Moore at Intel (as a start-up), bringing revenues to $20.8 billion in 1997. He was a strong manager and technologist, and respected as both.

  Dr. Grove was chosen as *Time Magazine*'s "Man of the Year" in 1997. He was also known as "The Father of OKR" (Objectives and Key Results) in the field of management.

- **Steven Jobs**. After difficult times, Steven Jobs brought Apple from the brink of disaster to the highest cap value in the country, some $337 billion (in 2011). Although Jobs was not an engineer, he deeply understood design trade-offs which is a critical skill of the systems engineer. These trade-offs often led to superior design in Apple's line of products. He had his share of failures but had the perseverance to move forward despite their toll on him.

Jobs was an iconic and charismatic character whose company led an industry with its computers and electronics, to include the iPhone, the iPod, and the iPad.

- **Jack Welch.** In 1981, as a young PhD engineer, Jack Welch took over as CEO of GE and 21 years later had increased their market capital by more than $450 billion. He understood how to "bash bureaucracy". Later, with his wife Suzy, he provided business consulting services using their treatise they've called confronting 74 of the toughest questions in business today. Here are some areas covered by their book: management principles, energy, continuous improvement, building trust, motivation, and innovation.

  Welch was named "Manager of the Century" by *Forbes* magazine and received the largest severance pay up (from GE) to that point, estimated to be some 420 million.

- **Simon Ramo**. Few do not recognize the "R" in TRW as Simon Ramo. He won many awards as an engineer and businessman, and had his name on some for others to receive. He lived a long life contributing to numerous engineering causes and being a super-mentor to many. He led the development of this country's microwave and missile technology. He possessed a wide-ranging intellect, as is true with many of the super-systems engineers.

  In addition, Dr. Ramo received many awards, to include the Pioneer Award from INCOSE, the Medal of Honor from the EIA, the Kagan Medal, the Founder's Medal from the IEEE, the Durand Medal, the Space and Missile Pioneer Award from the Air Force, and the NASA Public Service Award.

ENGINEERS FROM SPECIALTY BUSINESSES AND LABS

- **Benjamin Franklin**. Known to all and everywhere, Franklin was a leading scientist, publisher, diplomat, and inventor during revolutionary times. He is credited with having invented the bifocals, the lightning rod, and the Franklin stove. He was a major force behind establishing the University of Pennsylvania.

Ben Franklin served as postmaster-general for the British Colonies, governor of Pennsylvania, and Ambassador to France. He made scientific advances in several fields including the theory of light and electricity, meteorology, conductivity, and ocean currents, and published *Poor Richard's Almanack.*

- **Thomas Edison**. He was known as the Wizard of Menlo Park (New Jersey), very prolific with more than 1000 patents to his name. He is credited with systems involving electric power generation, sound, the phonograph, and motion pictures. He is also cited for his quotation "genius is one percent inspiration and 99% perspiration".

Thomas Edison is given credit also for the microphone, the alkaline storage battery, the Kinetograph (camera), the electrical vote recorder, and the ticker tape telegraph.

- **Kelly Johnson**. He established the "skunk works" at Lockheed where he operated just about independently under a set of rules that included (1) small project offices from both industry and the military; (2) it's important to minimize the number of reports, except for those that are important; (3) the project must have a limited number of very talented people; (4) project cost estimation and control is critical; and (5) he reports to a division president or higher. He is almost a company within a company with the autonomy usually afforded to a trusted subcontractor.

Kelly is known for some special 14 rules and practices, unique to the industry.

- **Tim Brown**. Tim Brown is the purveyor of "design thinking", a new way of approaching system design with special attention to systems thinking, visual thinking, smart teams, prototyping, and the rejection of prior negatives such as groupthink.

Tim Brown wrote the definitive text on design thinking, which had the name "Change by Design". He demonstrated how to integrate the key concepts of desirability, viability, and feasibility, as well as ideation, inspiration, and implementation.

- **Alexander Kossiakoff**. Alex Kossiakoff was born in St. Petersburg (Russia), studied chemistry at CalTech, and moved on to Johns Hopkins where he got his doctorate in 1938. He joined the Applied Physics Lab in 1946, pioneered in solid state fuel missiles, and directed the overall Lab's activities from 1969 to 1980. He was a Fellow of INCOSE and wrote a book on systems engineering with a co-author from the JHU/APL.

Dr. Kossiakoff received the Johns Hopkins President's medal and the DoD (Department of Defense) Medal for distinguished public service. The Kossiakoff Center at the Laboratory was named after him as was the Education Center.

- **J. Edwards Deming** was a master statistician, specializing in quality control and assurance. He gained prominence in that field from the work he did with the Japanese, resulting in significant increases in quality products. He served as an independent consultant in both Japan and the United States for many years, stressing 14 points that needed to be observed.

  Dr. Deming had clear messages about quality control and assurance, and delivered them in a series of seminars in the United States. He had thousands of devoted followers that carried forth his 14-point program for improvement.

- **Mervin Kelly**. Kelly was not extremely well known, but he served as president of Bell Labs and was significant in many of the Labs' contributions, such as the transistor, space communications, and information theory (from Claude Shannon).

  Dr. Kelly also served as a consultant to the Air Force, and a statesman for technology, Research and Development (R & D), innovation, and how to achieve creativity in a large organization.

  Apologies if you object to people with training and backgrounds in other than pure engineering (e.g., Deming) – there's clearly a method to this madness. I am presenting a "thumbnail" sketch for each of the above. A more detailed look by the author either reinforces, or not, the selection of traits that help the systems engineer to be successful.

THE LOGIC OF THE SURVEY

So we proceed with circumlocutious logic that goes something like this:

- a. We think we know, more-or-less, who the super-systems engineers are and have been.
- b. With this base of knowledge, we try to articulate the attributes of these engineers.
- c. These attributes are the very same as those that have led to their successes.
- d. We accept the limitations of sample size, and accept the positive aspects of intuition and direct observation by members and Fellows of the INCOSE community.

WHAT DO WE MEAN BY "SUCCESSFUL"?

The systems engineer today often operates in the domain of a system "program manager" who has the overall responsibility for the system. For our purposes, we will call this person the system "overlord". This person is often an employee of the government, but can also be a subcontractor. The "overlord" is also identified here as the "chief stakeholder" for the system. The systems engineer is successful when the chief stakeholder says so, and is pleased by the work done by the systems engineer. Another indicator of success is when the chief stakeholder designates the systems engineer as such by a letter of recommendation. We will not attempt, in this book, to further define in detail what it means for the systems engineer to be successful.

TODAY AND YESTERDAY

We note that there are differences between systems engineering in the 1500s and systems engineering in the 2020s. So it seems evident that today's systems engineer faces different challenges and environments than did the systems engineer of 600 years ago. However, we rely upon the good sense of today's outstanding systems engineer to see into and through fundamental skills and behaviors leading to our answers for today.

WHAT ARE THE ANSWERS FOR TODAY?

The seven attributes of today's successful systems engineer are taken to be:

    a. Synthesizer
    b. Listener
    c. Curious/systems thinker
    d. Manager/leader
    e. Expert in systems engineering processes
    f. Expert in domain knowledge
    g. Perseverer.

These answers are derived from a look at a variety of super-systems engineers, what they did, and how they did what they did and what they wrote about. Brief comments complete this short chapter.

a. **Synthesizer.** The very essence of systems engineering activity is to design and build systems that did not exist previously. This is fundamentally a synthesis process, as contrasted with an analysis process.

b. **Listener.** We are concluding that the independent skill of listening needs to be elevated to a distinct success factor. Without it, the activity is almost certainly to fail.

c. **Curious/Systems Thinker.** These two go together in often not very obvious ways. After being drawn into the problem and phenomenon by a basic curiosity, one or more "thinking" mechanisms come into play. [3] We presume that for systems engineers, a "systems" approach is appropriate.

d. **Manager/Leader.** Today's successful systems engineer must be able to manage and/or lead a team of very competent engineers. This is a non-trivial task and requires a definitive set of special skills.

e. **Expert in Systems Engineering Processes.** By and large, these are the processes of OSI/IEC 15288 as discussed in some detail in the *INCOSE Systems Engineering Handbook* [4]. Happily, all that work was not done for nothing; it's a real contribution.

f. **Expert in Domain of System.** Deep knowledge in the domain of the system being built is essential to success. You can be helpful with process expertise, but you will still miss the target without deep system domain knowledge

g. **Perseverer.** This personal characteristic is an integral part of the equation. It also shows up here as part of the success attributes. Take a good look at the book as well as the movies on "grit". And don't forget to apply the eyepatch.

REFERENCES

1. Gelb, M. "How to Think Like Leonardo da Vinci: Seven Steps to Genius Every Day", Dell Trade, 1998.
2. Thorpe, S., "How to Think Like Einstein", SourceBooks, 2000.
3. Eisner, H., "Thinking – A Guide to Systems Engineering Problem-Solving", CRC Press, 2019.
4. Walden, D., "INCOSE Systems Engineering Handbook, Fourth Edition", John Wiley, 2015.

Synthesizer

3

The first attribute on our list of systems engineering attributes is "synthesis". Recall that this is a non-ranked list, but synthesis is well-placed and well-conceived. We take the position that bringing a system into being from a clean sheet of paper is a most-valued activity and essentially the most difficult of the tasks of the systems engineer. And a first step at synthesizing has to be systems architecting.

OVERVIEW OF DODAF AND SELECTED DETAIL

We recall two distinct procedures for systems architecting. The first is the DoD approach [1] known as DoDAF. The essence of that approach is the notion of three views, namely,

1. The operational view
2. The systems view
3. The technical view.

The top-level breakdown of these three views gives us the following viewpoints:

1. All Viewpoint (AV)
2. Capability Viewpoint (CV)
3. Data and Information Viewpoint (DIV)
4. Operational Viewpoint (OV)
5. Project Viewpoint (PV)
6. Services Viewpoint (SvcV)
7. Standards Viewpoint (StdV)
8. Systems Viewpoint (SV).

The views have essential and supporting *products* as follows:

   a. Essential Views (with annotations)
     AV-1 Overview and Summary Information
     AV-2 Integrated Dictionary
     OV-1 High-Level Operational Graphic Concept
     OV-2 Operational Node Connectivity Description
     OV-3 Operational Information Exchange Matrix
     SV-1 System Interface Description
     TV-1V Technical Architecture Profile

   b. Supporting Views (without annotation)
     OV-4 Command Relationship Chart
     OV-5 Activity Model
     OV-6a Operational Rules Model
     OV-6b Operational State Transition Description
     OV-7 Logical Data Model
     SV-2 Systems Communication Description
     SV-3 Systems Matrix
     SV-4 System Functionality Description
     SV-5 Operational Activity to System Function Traceability Matrix
     SV-6 System Information Exchange Matrix
     SV-7 System Performance Parameters Matrix
     SV-8 System Evolution Description
     SV-9 System Technology Forecast
     SV-10a System Rules Model
     SV-10b System State Transition Description
     SV-10c System Event/Trace Description
     SV-11 Physical Data Model
     TV-2 Standards Technology Forecast

We see a very strong architecting foundation with these quite specific essential and supporting views. At the same time, it is not completely clear as to how to construct some of these views. What is clear, however, is that the systems engineer as architect is a critical role and capability.

ARCHITECTING USING THE EISNER ARCHITECTING METHOD (EAM)

The other approach to architecting is the construction by this author [2] which is set forth as a cost-effectiveness issue, consisting of the following four steps:

1. Functional decomposition
2. Synthesis
3. Analysis
4. Cost-effectiveness evaluation.

We take note of the fact that "synthesis" is step 2 in the above process and also that functional decomposition is the step before synthesis. So if we are looking at synthesis in general, we can usually expect functional decomposition as an earlier step. That also means that if we fail to do functional decomposition correctly, the error will tend to propagate into whatever it is that we are synthesizing.

Implicit in the synthesis step is the construction of several alternative architectures. This means that the analysis step is an analysis of alternatives (AoA). The body of knowledge about AoA can therefore be brought into play at that point. Finally, step 4 leads to the selection of a preferred architecture.

THE NOTION OF ALTERNATIVE ARCHITECTURES AS PART OF SYNTHESIS

As part of the synthesis step, more than one architecture is formulated. This is a natural result of a combinatorial examination of design choices. We also note here the fact that Einstein, in his approach to problem-solving, was fond of looking at combinatorics. Let us take another look at this issue from the point of view of system functions for a communications system, as shown in Figure 3.1.

System Functions	Approaches to Instantiate Functions
1. Modulation-Demodulation	A11, A12
2 .Switching and Routing	A21, A22, A23
3. Multiplexing-Demultiplexing	A31, A32
4. Encryption-Decryption	A41, A42, A43
5. Formatting/Signal Conversion	A51, A52
6. Control and Monitoring	A61, A62
7. Recording and Playback	A71, A72
8. Satellite/Terrestrial Communication	A81, A82

FIGURE 3.1 System functions for a communications system.

The array of approaches to the various functions, in principle, can generate as many as $(2)(3)(2)(3)(2)(2)(2)(2) = 576$ alternative architectures. A much smaller number is obtained from discussion by the architecting team as well as a search for a knee-of-the-curve solution. This is an important step toward providing a specific way to carry out the process of synthesis.

INTEGRATION

We do not have to look very far to find another area in which it is necessary to be skilled at synthesizing. The name of that area is simply "integration". This is accepted as a valid task of the systems engineer, indeed more than valid. It is crucial. We find it squarely in place in the ISO/IEC 15288 [3] and also in the real world that is inhabited by the systems engineering team.

The integration process is cited explicitly as part of the technical processes, which consist of the following:

a. Stakeholder requirements definition processes
b. Requirements analysis process
c. Architectural design process
d. Implementation process
e. Integration process
f. Verification process
g. Transition process
h. Validation process
i. Operation process
j. Maintenance process
k. Disposal process.

To assist in the task of integration, we offer here some specific suggestions, taken largely from an earlier treatise from the same author [4].

1. When attempting to integrate stovepipes, be careful so as not to promise anything close to 100% integration. Some stovepipes were designed for downstream integration but most were not. In this latter case, trying to crunch the stovepipes together will be difficult and very expensive both in time and dollars. Many stovepipes are "out there", looking for a way to seamlessly meet other integrable stovepipes.

2. Demand that all members of the integration team have definitive backgrounds in both project management and systems engineering. INCOSE certification is also a good place to look for qualified engineers.
3. In the "integration world", treat requirements as tentative needs that can be variables if and when the various stakeholders wish to do so. This can be a blessing in disguise.
4. Accept technology insertion as a desirable downstream activity that may enhance the overall system design and also create integration problems and issues
5. Always try to architect a set of alternatives from which to select the preferred alternative. See if it is possible to use the DoD Analysis of Alternatives (AoA) methodology to gain extra leverage and credibility
6. Assure that all systems are subject to a serious risk assessment and mitigation analysis.
7. Confirm that the ultimate system has sufficient schedule and budget to build, test, and install.
8. Reduce complexity and implement the K.I.S.S. principle wherever possible.
9. Understand and accept the acquisition ground-rules under which the system is being acquired. Make sure that a single person is charged with this responsibility, including the dissemination of information to members of the integration team.
10. Attempt to gain efficiency and leverage by means of reuse methods while avoiding the pitfalls of reinventing the wheel. When systems are being integrated, there are opportunities for reuse, but there is also more uncertainty that one is dealing with.

SYNTHESIS ACCORDING TO HALL

C. Hall [5] has set forth the proposition that the 20th century is the 100 years of analysis (dominated by analysis) and that the 20th century is the age of synthesis. He explores several reasons for this transition, to include improvements in thinking and technology. However, whatever the reasons, if we accept the premise, engineers are quite likely to play an important role during both the 20th and 21st centuries. Further, these trends will show up in the arts and sciences, as well as in what he calls the learned professions (including engineering) and the "applied fields". As part of his overall thesis, he has a quote from Emerson, as follows:

> the analytic process is cold and bereaving, and shall I say it…the poet sees the wholes and avoids analysis

A CHARMING QUOTE, IS IT NOT?

Hall comments on Peter Drucker who is the leading business consultant and researcher of his day. He also confirms that Deming would be a force in the direction of synthesis (new products) and Drucker looking at new ways to synthesize the products and processes of business. I have some dozen of Drucker's books on my bookshelves and I am not disappointed with his thinking, from both points of view (i.e., analysis and synthesis).

SYNTHESIS ACCORDING TO THE AVIATION ADVISORY COMMISSION (AAC)

Some years ago, an Aviation Advisory Commission was established, running for two years, to explore what national aviation should look like in the future. In other words, if we know where we should be going in, say 50 years, we might do a better job of preparing for that future.

The Executive Director for the Commission did a masterful job at synthesizing alternative futures for the consideration of the Commissioners. This was a distinctly synthesis task since he was suggesting designs and configurations that did not exist at that time. The alternatives were considered in terms of the following [6]:

Concept A – Extension of Current Operation. For each concept, there was a close look at and synthesis of the postulated system, to include operational notions, airport configurations, air vehicles, airspace consideration, ground access, and typical trips. The ground access segment typically had highway, rail, and air components.

Concept B – High-Density Short-Haul Supplement. Same as Concept A, except that a short haul supplement was added

Concept C – Remote Transfer Airport Supplement. Same as Concept A, except that a remote transfer airport supplement was added

Concept D – Local Terminal and Exchange Port Supplement. Same as concept A, except that two supplements were added: a set of local terminals and some number of exchange ports.

Here, we see a large-scale systems design (architecture) different from the classic single system configuration. This was a real challenge for the systems engineer – the Executive Director, who was shorthanded for this difficult task. The evaluation criteria for this system were established by the Executive Director as [7]

    a. International/Economic
    b. Social
    c. Environmental
    d. Quality of service
    e. System capacity
    f. Human factors
    g. Investment costs
    h. Operating costs.

Due to the size and difficulty of the task of evaluating four alternatives against eight criteria, a full quantitative analysis was never completed and documented. Here, we see a system that compares in size with the above national aviation system. However, it has a more focused purpose, i.e., to shoot down as set of threats (e.g., missiles) that are firing at your missile assets, whether they be ship-based or ground-based.

A considerable amount of attention has gone into the command and control portion of this national defense system. This is an example of another synthesis task that faces the systems engineer in the next several years.

SYNTHESIS OF AN AIR DEFENSE SYSTEM

The very large air defense system will have to be architected, as per the overall procedures for systems architecting. The process started in about 1981 under President Reagan and the Strategic Defense Initiative (SDI) program. Here, we see a system that compares in size with the above national aviation system. However, it has a more focused purpose, i.e., to shoot down a set of missiles that are being fired at you.

This is a quite complex scenario with a constellation of satellites and some number of weapons aboard each satellite. Variables of interest in this scenario include

a. The number of target engagements
b. The dwell time
c. The number of weapons engaging targets
d. The total time available
e. The cost per satellite
f. The absentee ratio.

After some deliberation, the relationships between these and other variables can be constructed. But it is still "a long way from home". And this is largely the analysis part of synthesizing a complex system.

PUTTING THE PIECES TOGETHER

As noted previously, the systems engineer must be able to "put the pieces together", otherwise known as synthesize. The key situation in terms of what the pieces are and where they came from is that of architecting a system. So in that sense, the synthesis is operating as system architect. There are, however, situations where synthesis is called for, but not in terms of the architecture of a system. Generically, we would simply call them decision meetings. These are meetings at which lots of data are presented and the agenda calls for decisions regarding what all of the data means. The systems engineer is being called upon to put the pieces together so as to facilitate the decision-making of the group. This is a very important situation, and the systems engineer plays a key role in it.

A SIMPLE EXAMPLE – SYNTHESIS AS A STEP IN THE SYSTEMS ARCHITECTING PROCESS

We gain further insight into the matter of synthesis by considering it as a part of the architecting process (EAM). We recall the first step in the process as that of functional decomposition. We select a relatively simple information technology (IT) system as our example and have the following illustrative decomposition.

FUNCTION LEVEL 0	THE OVERALL SYSTEM (LESS POWER SUPPLY)		
FUNCTION LEVEL 1	SYSTEM A	SYSTEM B	SYSTEM C
1.1 Input	Mouse, keyboard	Add screen tap	Add voice
1.2 Output	Laser printer	High-speed printer	Multicolor printer
1.3 Processing			
1.4 Cyber	Low-end Norton	Mid-range Norton	High-end Norton
1.5 Database	Access	Mid-range Oracle	High-end Oracle
1.6 Apps	Five function App package	Add another five functions	Add another ten functions

We note how this is illustrated. In particular, we have three alternatives for each function and sub-function. Design alternatives selected with great simplicity to demonstrate the point and format. The "analysis" step follows. In that step, we evaluate (rate) each of the three alternatives against a set of criteria, each of which has an assigned weight. We calculate the products (w x r) for each alternative and then sum. That summation is represented as a set of MOEs (measures of effectiveness). From that point, we move into classic cost-effectiveness evaluation leading to a selection of the preferred system architecture. An AoA approach as defined by the Department of Defense [7] should also be considered.

EVALUATION CRITERIA	WEIGHTS	SYSTEM A RATING WEIGHT TIMES RATING (WXR)		SYSTEM B RATING WXR		SYSTEM C RATING WXR	
Speed	0.3	4	1.2	6	1.8	9	2.7
Size/capacity	0.2	5	1.0	6	1.2	9	1.8
Maintain	0.3	4	1.2	5	1.5	7	2.1
Overall performance	0.2	3	.6	7	1.4	9	1.8
TOTALS	1.0		4.0		5.9		8.4

If we add the cost estimates for the three alternatives, we obtain

	SYSTEM A	SYSTEM B	SYSTEM C
Effectiveness	4.0	5.9	8.4
Cost	$3.0 m	$5.0 m	$10 m

Typically, these two values would be plotted on an x-y (cost-effectiveness) graph to visually see how the three curves compare.

THE BOTTOM LINE

The bottom line is that a key attribute for the systems engineer in order for he or she to be successful is the ability to synthesize. This is often translated into the position of chief engineer. It is also often connected to one or more quite specific projects which raises the question: Does the chief engineer "work for" the project manager or does the project manager work for the chief engineer? The answer is simply that as strong as the chief engineer might be in all the important dimensions (e.g., years' experience and degrees), he or she reports to the project manager on all matters related to that project.

If we look at the negative side of the proposition, we ask ourselves this question: if the systems engineer fails to have the attribute of the ability to synthesize, does this mean that he or she is quite likely to fail? This author would answer "yes" to that question. Perhaps the reader would be more lenient and answer – "it depends". For this author, it does not depend; the systems engineer *must* have the critical skill of being able to synthesize.

REFERENCES

1. DoDAF 2.02, Department of Defense, CIO.
2. Eisner, H., "Systems Architecting – Methods and Examples", CRC Press, 2020.
3. ISO/IEC 15288, "Systems Engineering – System Life Cycle Process", 2002.
4. Eisner. H., "Essentials of Project and Systems Engineering Management", 3rd Edition, John Wiley, 2008.
5. Hall, C., "The Age of Synthesis – A Treatise and Sourcebook", Peter Lang Publishers, 1995.
6. Eisner, H., "Computer-Aided Systems Engineering", Prentice-Hall, 1988.
7. "Analysis of Alternatives Handbook", see www.prim.osd.mil.

Listener

4

We move on next to another critical skill that makes a systems engineer successful, or not. This skill has to do with "listening", in the full meaning of the word and action. In this connection, we envision that the systems engineer is fully engaged with a member of his team, listening to every word and acknowledging receipt and understanding. When was the last time you had this experience with another engineer? When was the last time you noticed that your colleague, an engineer, was not really listening to what you were saying to him? When was the last time you felt completely in touch – in the sense that your message transmitted was truly received and understood? That is the skill that we are referring to in this chapter and with respect to the important capability of the systems engineer.

THE CHALLENGER ACCIDENT

The shuttle Challenger accident occurred on January 28, 1986. It led to an immediate listen-in at NASA to find out what happened. NASA lost no time in setting up hearings, understanding the incident could be a life-and-death issue for the agency.

NASA's head at the time was William Graham who picked up the phone and invited Prof. Richard Feynman to participate in the inquiry [1], which ultimately took the form of a commission under a gentleman by the name of Rogers. Feynman accepted the invitation to listen-in, with no pre-judgment as to where all that would take him. He would listen first and then make up his mind, based upon what he heard. First, he would listen.

Key NASA engineers and managers moved ahead with briefings to the members of the commission. Feynman was mostly in a listening mode, but asked questions when something was missing or not clear. The commission was allocated 120 days to complete their task.

After a while, Feynman zeroed in on a set of O-rings as the possible problem, leading to the overall catastrophic failure. Feynman not only listened hard, he set forth a brief experiment to try to show what was wrong with the O-rings. He did this by immersing an O-ring in ice water showing that the ring did not exhibit resilience when the water temperature was below a certain value. That was, in effect, a design defect that was visible to all. And Feynman got a lot of credit for "solving" the problem and doing it in such a novel way. If we look carefully at what happened, we find that it was all made possible by Feynman's keen ability to listen, and draw conclusions in a field that was basically foreign to him. After all, he was a physicist, not an engineer. But he knew how to listen, and to follow up with the right steps.

CORPORATE LISTENING

We are finding that more and more attention is being paid to communications between engineering groups. This overall communications, it has been shown, increases productivity as well as creativity. And it goes beyond discussions at the water cooler, facilitated by proximity. It goes into communities of interest, meetings, special task forces, and other forms of communications. All of these, of course, cost money and time, and reflect an expanded corporate commitment. They demonstrate a further acknowledgment of value and return on investment (ROI).

Increasing communications has a built-in assumption of more transmitting and more receiving, which leads, of course, to more listening. We measure the listening where we can, but otherwise take it for granted or see it in small acts by our employees. Just as Watson at IBM changed the corporate culture with his "Think" Notepad, many CEOs are changing their corporate cultures by showing how listening can be carried out and how more of it has its distinct benefits.

Here's an example. Some years ago many companies experienced pressure to improve the quality of their products and services. The cultures had to change, and the CEOs knew they had to lead the charge. Quality circles appeared everywhere, and everywhere, people were listening to PowerPoint lectures on how the company was going to do some very new things with respect to quality. Some companies even hired Dr. Deming, the "Father" of Quality Assurance to lead them down this new pathway. I can still remember (I listened hard) the presentations that I attended, given by the indomitable Dr. Deming. Listening was difficult since he did not project well, and he also mumbled a lot.

Another example may be gleaned from the Hewlett-Packard (HP) Company. First, they had an open-door policy which tended to assure a culture of listening.

Second, in their book [2], they affirmed their policy of "listening to customers". Third, in that same book they confirmed their commitment to participatory management. That too was a statement, if you will, of listening and culture. This extraordinary company always makes me smile when I think about who they are and their approach to how they built the company. Above all, in terms of an enduring culture, is that they trust their people. That results in superior performance and a huge set of employees that trust the company. The company may make some mistakes from time to time, but the trust is still there between employee and management. And above all, they're listening to each other.

ACTIVE LISTENING

There is a type of listening that has been called active listening [3]. In such situations, the listener is "actively engaged" with the speaker, leading to better overall results. Signs of active listening are

a. Interesting facial expressions
b. Some degree of gesticulating
c. Some amount of smiling and frowning
d. Asking and answering short questions
e. Other forms of nonverbal communication.

According to the Conover Company [3], up to 93% of communication is nonverbal. So we all have to pay attention to what's going on beyond the words. Perhaps we need also to remember the Marshall McLuhan suggestion:

• The medium is the message.

McLuhan was saying – at times, there's a lot going on beyond the physical message. There's the medium, and that too can be part of the message. If it is, it's part of listening.

THE ART OF LISTENING

Google searches in pursuit of material for this book led to TED talks about the art of listening. As it turned out, there are quite a few folks that have studied listening, from many angles. Some of the TED talk names included William Ury, Leon Berg, Jason Chare, Lalita Amos, and Mteto Maphoni. Not familiar names but all had something serious to say about listening. Tune in to your local TED talk person and find out. But beyond that, some just pointed out that listening includes not only hearing by the ear but also the next cognitive brain function that calls for interpretation of what was heard. What was heard could be complicated, or it could be exceedingly simple. It could be just plain tuning in, or it could be storage, retrieval, and search for meaning. People have been struck by what they just heard, as in – I wonder what he/she meant by that?

ONE WOMAN'S VIEWS

We cite here several views set forth by a well-known columnist and essayist, Terry Schmitz [4], who had special feelings about listening. First, when there is honest and heartfelt listening, there will also be a new flow of creativity. People appreciate the attention and feel closer to one another when they can experience the connection of listening. Ms. Ueland conveys a few instances when she felt that her elders were really listening to her, with full attention. She recalls the positive effect that had on her, and also recalled the negative side of not listening.

She was especially struck by an elderly patient's reaction to her:

> I had listened with such complete, absorbed, uncritical sympathy, without one flaw of boredom or impatience, that he now believed and trusted me.

Based upon her experience, the sensitive experience of an essayist, she flat out declared that "women listen better". What do we think about that? This author is not surprised but would like to hear more and see more evidence from experiments.

ONE MAN'S VIEW

Thomas Anderson, [5] from the Engineering Management Institute, makes the point that accurate communications for engineers are especially important. His point is well taken. Many systems put together by engineers have life and death elements or modes of operation. Think about bridges and subway systems. Mistakes can literally be fatal. In any case, Anderson leaves us with his list to clear communications:

  a. Listening
  b. Consistency of communications
  c. Patience
  d. Follow-up when necessary
  e. Clear and concise speech
  f. Repetition if any sign of confusion.

ONE DEAN'S VIEW

The founding dean at the Chapman University's new engineering school stressed listening as a skill that needed to be reinforced [6]. Dean Andres Lyon set up "listening sessions" to assure that all had a common vision as to what the school was about and how to establish a new culture. Faculty went over plans and value exercise to assure that all were on the same page. Emphasis was placed upon trying to make sure that everyone was "listening".

LISTENING AS ENGINEERING COMMUNICATION

Pioneers at the IEEE [7] have argued that listening is a key attribute for engineers, supported also by the National Academy of Engineering and the Accreditation Board for Engineering and Technology. The argument is further represented in terms of two types of listening:

  a. Basic listening
  b. Contextual listening.

The former is what we normally mean by speaking and hearing. The latter depends upon other context factors that might facilitate deeper meaning. The authors suggest that contextual listening may well be a better support for design and problem-solving activities on the part of engineers. The ways in which this may be achieved, according to Leydens and Lucena, are

a. To develop a two-way (or more) empathic relationship between communicators
b. To transform actions into ones that are collaborative, with stakeholder and project ownership.

The bottom line here is that the above authors have been champions of listening by engineers and have spent time exploring how it supports a deeper meaning and understanding between listeners and speakers. This author agrees as to the importance of listening, which is the reason it has appeared on the overall list of success factors for the systems engineer.

COMMUNITIES OF PRACTICE

Some years ago, the company org chart had a blip called security. That blip soon stood for both physical security and information security (infosec). The old IT department had to worry about infosec, and all of a sudden, there were some new things to worry about. People in the company enthusiastically formed a "community of practice" (CoP) so that infosec issues could be handled with excitement and efficiency. As I remember it, there were many volunteers to serve on the infosec CoP, and they were not disappointed. New domain, new people, new threat, new technology, new challenges, and new problems to solve.

You learned a lot by being part of it and also by listening to your colleagues that served on the infosec CoP. Yes, it had a large listening component, and you became aware of the need to improve your listening skills. Was there too much to listen to, too much to absorb? The answer was "yes", but engineers, by and large, like to take on this type of new challenge. By way of example, here are some of the issues that an infosec CoP might have to confront:

a. Intrusion by a third and unknown party
b. Use of special keys
c. The use of special crypto technology
d. New levels of classification
e. New badges for all employees
f. Evaluation of infosec software
g. Internal and external compatibility and interoperability
h. Transfer of classified material between facilities.

DOES NOT LISTENING LEAD TO GROUPTHINK?

The previous chapter introduced the notion of GroupThink in terms of dysfunctional group behavior. Here, we look at listening and ask the question: is it likely that if there is no listening, the likelihood of GroupThink is increased? This author believes that the basic answer to that question is "yes", although I have not found corroboration in the literature. Perhaps readers have some relevant experience in this specific area and can send an e-mail to me.

ONE-ON-ONE NOT LISTENING, AND LISTENING

You, the reader, have very likely been party to one-on-one not listening. It certainly is not pleasant to see your partner out in Neverland when you are talking to him or her. Aside from being rude, this behavior leads to no progress on the tasks before you. If you can tune in to this experience, take a chance and bring your partner to a hearing and listening mode of behavior. This may well solve a real problem that your partner has had for a long time.

So imagine that you're sitting across the table from your work partner, having a conversation. Your partner exhibits the following behavior, suggesting that he or she is listening:

    a. Eyelids are up, and not closed.
    b. An intense stare at you.
    c. Lips are starting to move; listening has terminated.
    d. You see a smiling face.
    e. The face you see lights up with comprehension,

And assuming you're listening:

    a. You feel yourself smiling as a thought is received.
    b. You re-state the meaning of what you just heard.
    c. You re-phrase what you just heard, and verbalize.
    d. "I hear what you're saying", you say by way of feedback.

WHAT DOES *FORBES* HAVE TO SAY?

So an article in *Forbes* [8] addresses the matter of listening quite directly. Here are five suggestions, boiled down from the original ten. These tend to stress the contact, one way or another, between the speaker and the listener:

1. Try to make contact (eye, question, etc.) with the speaker.
2. Do (1) above but without interrupting.
3. Be obviously attentive.
4. Be open to new ideas and new ways of expressing them.
5. Try to provide feedback to the speaker.

Now we have a quick note from the "Father" of Management [9], Peter Drucker:

As the first basic (leadership) competency, I would put listening. It is not a skill; it's a discipline. Anybody can do it. All you have to do is keep your mouth shut

Listen not to confirm what you believe to be true, but to explore what else is true

Learn to listen, not to respond

And some additional advice – "all you have to do is tell people". "Look, I'm a listener. Before you give me that darn report, tell me what's in it" [10].

A BOTTOM LINE

The bottom line for this chapter is that "listening" has been established as a success attribute for the systems engineer. That is, if he or she has this attribute, they are more likely to be successful as a working systems engineer. This author also goes so far as to suggest that if this person does not listen, most of the time, he or she is not likely to be successful.

REFERENCES

1. Feynman, R., "What Do You Care What Other People Think?" W & W Norton, 1988.
2. See https//engineeringmanagementinstitute.org/communication-engineers-important.
3. Packard, D., "The HP Way", HarperBusiness, 1995.
4. Schmitz, T., See www.conovercompany.com/teamwork-active-listening.
5. Brenda, U., "Strength to Your Sword Arm", Holy Cow! Press, Duluth, MN.
6. Leydens, J. and Lucena, J., "Listening as a Missing Dimensions in Engineering Education: Implications for Sustainable Community Development Efforts", *IEEE Transactions on Professional Communication*. Vol. 52, no. 4, pp. 359–376, 2009.
7. Lyon, A., "Listening Sessions", *PRISM Magazine*, American Society for Engineering Education, March–April 2020.
8. Schilling, D., "Ten Steps to Effective Listening", *Forbes Magazine*, 11/09/2012om/peter-drucker.
9. See https://keithwebb.com /peter-drucker.
10. Drucker, P., "Managing Oneself", The Drucker Lectures, McGraw-Hill, 2010.

REFERENCES

Curious/ Systems Thinker

5

We start this chapter with a brief exposition of quotes, which seem to be in abundance.

> Curiosity is one of the permanent and certain characteristics of a rigorous intellect.
>
> *– Samuel Johnson*

> The World is but a school of inquiry.
>
> *– Montaigne*

> Life must be lived and curiosity kept alive. One must never, for whatever reason, turn his back on life.
>
> *– Eleanor Roosevelt*

> It is a miracle that curiosity survives formal education.
>
> *– Einstein*

> Curiosity killed the cat, but for a while, I was a suspect.
>
> *– Steven Wright*

> The first and simplest emotion which we discover in the human mind, is curiosity.
>
> *– Edmund Burke*

We refer back to the first chapter and note that da Vinci had an "insatiable curiosity", as did many other super-systems engineers. All these engineers "suffered" from an overwhelming curiosity that helped them with their reading and their investigation and their perseverance. A wonderful affliction, if you will – a powerful and ever-present curiosity. If only there were enough time every day to follow up with responding to the roads that we are led to from this sense of curiosity.

The reader is reminded that we are in search of success attributes of the super-systems engineer. Curiosity and thinking is one of the suggested main attributes, supported by the discussion of this chapter. We will not explore the negative side of the hidden question: will the systems engineer fail if he or she does not exhibit, and does not possess, this "curiosity" trait.

Curiosity will lead us to asking, and eventually answering, difficult questions. We will attempt to illustrate this point by setting forth several key questions to be addressed in these next several years.

1. **General Systems Theory**. Several researchers (Bertalanffy [1] and Forrester [2]) have tried to tackle this issue. Some success has been achieved. Can we proceed with a more complete and useful "systems" theory?

2. **Ubiquitous MBSE**. Is it possible to move ahead and formulate a generalized model for Model-Based Systems Engineering (MBSE)? Perhaps this would look a lot like what Jay Forrester did with System Dynamics?

3. **Software and Systems Architecture**. Although one can find literature (books, for example) on how to architect a software system, and how to relate that construct to a systems architecture, can we move further down that road in this domain?

4. **National Aviation System (NAS) Model**. Moving into domain-specific systems and issues, we need an overview model that is capable of handling the entire NAS. Do we have an idea as to how to proceed in this direction?

5. **Air Defense Model**. Another domain of special interest is that of air defense. This applies to the overall "national defense" issue, a derivative of the 1980s Strategic Defense Program (SDI) and moved ahead under the name "National Missile Defense".

6. **International Negotiation Model**. We would appear to be in great need of this type of model to help us deal with treaties with both friendly (e.g., Mexico, Canada) and unfriendly (Iran, Russia) countries.

Are our best systems engineers interested in (curious about) these questions and their answers?

By way of getting closer to serious thinking on the NAS model, we find that it is not difficult to define a dozen or so key variables for such a model.

The major difficulty, of course, is to formulate meaningful relationships between all these variables, which are listed below:

Key Variables for a NAS Model (*)

1. Trip Time
2. Speed
3. Frequency of Service
4. Passenger Capacity
5. Freight Capacity
6. Turnaround Time
7. Demand
8. Capacity/Demand Ratio
9. Airport Capacity
10. Delay
11. Noise
12. Pollution.

(*) aggregation by source, local/national/international

THE LEMELSON CENTER

In an article entitled "Curiosity and Invention" [3,4], the question "Do you consider yourself curious?" is set forth. The Lemelson Center, for a couple of decades, has been studying curiosity and innovation on the part of individuals and institutions. One conclusion is that "inventors are insatiable and unstoppable in their curiosity and their quest for improvements". Another conclusion is simply that "curiosity is key to the invention of products and systems, but in a sustained, disciplined, and programmatic way". Of course, our patent system is set up to be rigorous and competitive, with limits on the rate at which inventions are processed. So – one might say, if you're not overtly curious, you'll have to wait at the end of the line.

The Center states as its mission the following:

The Lemelson Center engages, educates, and empowers the public to participate in technological, economic, and social change. We undertake historical research, develop educational initiatives, create exhibitions, and host public programming to advance new perspectives on invention and innovation and to foster interactions between the public and inventors.

The Center is part of the Smithsonian and is located at the National Museum of American History.

STRATEGIC PLAN OF CENTER

Items that are cited as part of the Lemelson Center strategic plan include

    a. Study of invention and innovation
    b. Empower public to solve problems
    c. Effect change
    d. Advance scholarship
    e. Accept and navigate challenges
    f. Share stories
    g. Nurture creativity in young people.

The Center's location in terms of the Museum brings more people to its entrance and facilitates interactions dealing with invention and innovation.

A CONSEQUENCE OF CURIOSITY

The natural flow from a deep curiosity is more thinking about more and more subjects. So our "thinking gland" is stimulated, and all that is a good thing. Our curiosity and thinking are the "one-two" punch that represents a strong capability in our best systems engineers.

Curiosity leads people to take the next step, namely, thinking. This author explored various thinking approaches and processes in a book by that name [3]. "A bottom line" list shows the conclusion as to the diversity of approaches cited:

    1. Inductive thinking
    2. Deductive thinking
    3. Out-of-the-box thinking
    4. Reductionist (Descartes) thinking
    5. Systems thinking
    6. Design thinking

7. Disruptive thinking
8. Lateral thinking
9. Critical thinking
10. Hybrid thinking
11. Breakthrough thinking
12. Visual thinking
13. Fast and slow thinking.

This startling list is made even more eye-popping by the last item, developed in detail by the Israeli sociologist/economist Daniel Kahneman [5]. The basic idea behind his research and documentation thereof is that people, in general, have two modes of thinking: one is fast and the other slow. The former (system 1) is agile and kicks in quickly in response to a given arbitrary situation. This mode also is highly intuitive and emotional. The slower mode (system 2) is more deliberate and logical. There's more time to mull it over, to think more deeply about the given situation and proposition.

We assume that the above behavior is present in all of us, even as we are part of a team. The implications are that we might expect the system 1 behavior from many of us as issues are raised and as people are responding in real time to them. Possibly, that is a good name for the quick response system 1; the "real-time" response. Possibly, a closer re-read of Kahneman's book will reveal that he has already thought of this idea. A re-reading search has revealed a section that addresses what the author thinks might be the consequences of his theory, namely, that we would expect a lot of system 1 behavior at free-wheeling meetings. At such sessions, people tend to be given permission to say what it is that comes to mind, without a lot of censoring and intermediate time delays.

The Nobel-winning treatise and research from Kahneman was preceded by substantive work in what might be called psychological decision theory. In that domain, Kahneman worked with Amos Tversky to formulate the following propositions:

a. **Regression to the Mean**. This refers to the apparent fact that we have a tendency to let down after an excellent performance and pick up after a poor performance.
b. **Loss Avoidance**. Here, the behavior pattern is to avoid cost overruns and slippages in favor of risky decisions with some upside potential.
c. **Representativeness and Availability**. Both of these refer to establishing a mental model based upon prior experiences rather than using current facts. (Don't bother me with the facts, my mind is already made up!)

GROUPTHINK

This topic and phrase was set forth by a psychologist from Yale [6], and it refers to a particular mode of behavior when a group gets together and come up with a wrong solution due to the interaction between the people. That interaction causes various members of the group to agree to come to a consensus and not express contrary views. In other words, they don't agree with the consensus, but for various unspoken reasons, they do not wish to come out and disagree.

The "Bay of Pigs" incident is claimed by many to be a good example of GroupThink. Or, put another way, it's an example of the possible dynamic and conclusion when GroupThink is in effect. Apparently, many disagreed with the invasion solution but did not express their true thoughts, leading to a disaster. It is distinctly a bad arena (e.g., warfare) in which to allow GroupThink to occur. Other warfare examples have appeared in the literature (e.g., Vietnam, Iraq) reinforcing the point that it's now time for us to recognize and avoid GroupThink.

Another GroupThink example has been set forth with the name "The Abilene Paradox" [2], and by the George Washington University Professor J. Harvey. In his construction, there's a group of four that live in Coleman, Texas, that gets together and decides to drive to Abilene for a dinner. The weather is over 100°, and all they have is a non-air-conditioned Buick. Nonetheless, they take the trip, apparently feeling uncomfortable the whole time. When they return, they reveal to each other that they really didn't want to go Abilene, but they did it anyway. How do we account for this type of behavior? Prof. Harvey is able to write quite a bit about it, stressing that this is a quite serious problem that faces our civilization. That may or may not be confirmed, but the phenomenon is largely accepted as true.

A third example of GroupThink is the fact that the United States failed to intercept the Japanese invasion at Pearl Harbor. Apparently, we had all the information necessary to take pre-emptive action at that time. However, we did not do so, leading to severe personnel and equipment losses. Reason? Attributed to GroupThink.

A fourth example dealing with GroupThink suggests that the Vietnam War had many influential people against it, but they could not carry the day. So it's not just a decision exercise; it's also about the possible savings of lives. It is not difficult to see where and how GroupThink might occur in an engineering design setting or context.

THE SYSTEMS THINKER

It is suggested here that the systems thinker is a natural and short step beyond curiosity. Here are some of the features that appear to be important to the systems thinker [7]:

a. Holistic
b. Broad rather than narrow
c. Integrated
d. Expansive
e. Generalized
f. System-wide
g. Larger context
h. Lateral
i. Fusion
j. Top level
k. Inclusive.

Some years ago, Peter Senge wrote a blockbuster book [8] that addressed learning in today's corporations. He claimed that there were five parts to that puzzle – five disciplines. Four of the disciplines were

1. Personal mastery
2. Mental models
3. Team learning
4. Building a shared vision.

Then, he went on to fill in the fifth discipline – systems thinking. That's his answer to learning in a corporate setting – systems thinking. What we are saying here is that a key attribute for success of the systems engineer is one and the same – systems thinking.

A simple illustration of how a systems perspective helped in solving a problem was set forth by Barry Boehm [9], one of our very strong software engineers. It seems that some years ago, when Barry was managing an IT project, he ran into some not very wonderful news. They did a project assessment and found that they had an original budget for the system of $30 million, but now the calculation was that it would cost $100 million to complete. Barry set his team to work to define and explore the features of several alternatives. In this regard, they found that if they relaxed the response time requirement from 1 to 4 seconds, they could build the system for the original budget of

$30 million. After much analysis and gnashing of teeth, they decided to put this matter in front of the big boss, Barry's customer. After due deliberation, the boss made a declaration – let's go with the 4-second response time and the $30-million budget. We make the following two observations from this anecdote:

a. The problem was ultimately resolved when they realized that the customer was "part of the system" and a critical part no less.
b. A broader system look led to an exploration of alternative response times (which led to the resolution).

Thus, systems thinking is almost always looking at the problem from a broader perspective. New insights as to a possible solution are gained from such a perspective.

THE SYSTEMS APPROACH

This is the point of view taken by the systems engineer when starting out to build a large-scale system. There are variations on some of these themes, but the following define the overall approach [10]:

1. Establish and follow a systematic and repeatable process.
2. Assure interoperability and harmonious system operation.
3. Be dedicated to the consideration of alternatives.
4. Use iterations to refine and converge.
5. Create a robust and slow-die system.
6. Satisfy all agreed-upon user/customer requirements.
7. Provide a cost-effective solution.
8. Assure the system's sustainability.
9. Utilize advanced technology, at appropriate levels of risk.
10. Employ systems thinking.

REFERENCES

1. Bertalanffy, Ludwig von, "General System Theory", George Braziller, 1968.
2. Forrester, J.W., "Industrial Dynamics", Pegasus Communications, 1961.

3. Daemmrich, A., "Curiosity and Invention", The Lenelson Center; see: invention.si.edu/curiosity-and-invention.
4. Eisner, H., "Thinking – A Guide to Systems Engineering Problem-Solving", CRC Press, 2019.
5. Kahneman, D., "Thinking, Fast and Slow", Farrar, Straus and Giroux, 2011.
6. Janis, I., "Victims of GroupThink", *Psychology Today*, November 1971.
7. Harvey, J., "The Abilene Paradox" and Other Meditations on Management", Lexington Books, 1988.
8. Senge, P., "The Fifth Discipline – The Art and Practice of the Learning Organization", Doubleday, 1990.
9. Boehm, B., "Unifying Software and Systems Engineering", *Computer Magazine*, pp. 114–116, March 2000.
10. Eisner, H., "Topics in Systems", Mercury Learning and Information, 2013.

Manager/ Leader

6

There are many roles that a systems engineer plays, depending upon the arrangements that he or she has made. One role is to serve as a consultant to a stakeholder, performing whatever tasks are requested. These tasks can vary from individual performance to managing a typically small group of project personnel. Yet another role is to take on the job as project manager of a designated project. Let's start with the requirements of a project manager position.

MANAGER AS PROJECT LEADER

This is a classic position and has been documented in considerable detail in the literature [1] including work from this author [2]. The overall project has a simple triumvirate structure that consists of

1. The project manager
2. The chief systems engineer
3. The project controller.

The sub-structure of the project can have many functions having to do with such areas as

1. Hardware
2. Software
3. Architecting
4. Integration
5. Test
6. Verification and validation
7. Others that are specialized to the project.

47

Here is a "top five" citation of the core competencies of the project manager:

1. Planning, organizing, directing, and monitoring (PODM) skills
2. Problem-solver
3. Communicator, up and down the organization
4. Decision-maker
5. Representation of corporate culture.

PODM Skills. PODM skills are considered the "bread and butter" of project management. We need to know where we are at all times. There's a basic plan, and measurements are made against that plan. If we are off the plan, that triggers a review as to where we are.

Problem-Solver. Problems come across your desk all the time as project manager, and they need to be addressed, in real time. Many of the issues are not predictable, but they appear with regularity and certainty.

Communicator. Much of project management depends upon the project manager being able to communicate with all members of the team. At the same time, all members of the team are expected to be good communicators.

Decision-Maker. It is not enough to keep chewing on a variety of project issue. The project manager is expected to make decisions when called for, and also that they be good decisions.

Corporate Culture Representation. The project manager needs to interface between the corporate entity and the project so that both parties gain from the existence of the project.

The three dimensions of project management tend to be project cost, schedule, and performance. Issues involving one or more of these dimensions tend to dominate the concerns of the project manager. But in the final analysis, every project concern is a candidate for landing on the project manager's desk. He or she is perfectly justified in declaring "the buck stops here!".

The project manager has the advantage of being able to get some help from the chief systems engineer and the project controller. The former keeps his or her eye on system performance (are we meeting the system requirements?), and the latter helps with budget/cost and schedule.

Most engineering and business schools teach a course or two in project management. To illustrate, here is an overview of project management topics and required skills from academia [3]:

a. Ability to communicate
b. Function as a leader
c. Exhibit conflict resolution and negotiation abilities
d. Engage in strategic thinking
e. Extensive planning

f. Ability to organize

g. Targeted technical skills.

Communication. We tend to see this skill on just about every list.

Leadership. More about this later in this chapter.

Conflict Resolution. There are many arenas in which conflict can and will arise on a project. It is up to the project manager to resolve any and all conflicts.

Strategic Thinking. Many think that the domain of the project manager is limited to tactics. This is not the case. The project manager is called upon to do strategic thinking about where the project, and its derivatives, fits in the grand scheme of things.

Extensive Planning. The first order of business is the project plan. Then, as circumstances change, a revised plan is required. Many engineers have complained to me – all I do all day, every day, is planning, planning, planning.

Ability to Organize. The scope of this is obviously within the project, but it extends to interfaces with the organization at large.

Technical Skills. This is critical in terms of the performance of the system that is being built. Mismatches here will often lead to failure, e.g., a PhD nuclear engineer by original degree serving as project manager for a battlefield communication system.

UK PERSPECTIVE ON PROJECT MANAGER COMPETENCIES

For reasons that are not at all clear to this author, the UK perspective regarding (project manager) PM competencies differs from that in the United States. One source [4] defines the UK approach with the six abilities, which is described as follows:

Awareness of Profit. Translated, this means that the PM shall be aware of the need to make the profit that was planned and work toward that objective.

Technical Ability. Have technical skills that pertain to the overall business (i.e., computer, information technology, engineering, etc.).

Change Dynamics. Understand and accept changes that occur within and outside of a project. Suggest changes when appropriate.

Policies and Procedures. This pertains to the organization in which the project is located. Must contribute to and accept their policies and procedures.

Decision Theory. Deep in understanding of the theory behind posing alternatives and deciding how to make appropriate selections.

Duality. Able to balance objective and subjective factors and display integrative behavior. Some call this an appropriate synthesis between right brain and left brain.

One scans this list and is ready to declare that they were formulated on another planet. Possibly, that is actually the case.

THE UK PROJECT MANAGEMENT INSTITUTE (UKPMI)

If we investigate the Project Management Institute (PMI) in the United Kingdom, we find that they are much more aligned with the United States in terms of project management skills. Here is a list of such skills, obtained directly from the relevant website:

PMI skills as cited in the United Kingdom [5] are

1. Time management
2. Organization
3. Communication
4. Negotiation
5. Risk management
6. Leadership.

This citation of skills could easily have come from the United States.

PROJECT MANAGEMENT AND THE ISO/IEC 15288 STANDARD

We recall that there is a connection between what a manager does, in generic terms, and what a manager is called upon to do as a systems engineer. That connection can be identified as simply the standard known as ISO/IEC 15288 [6]. This standard calls out relevant "project planning processes", as below:

a. Project assessment process
b. Project control process

c. Decision-making process
d. Risk management process
e. Configuration management process
f. Information management process.

Beyond these processes under project planning, there are other processes of importance under technical processes, such as verification and validation.

SOME PROBLEMS FACING THE PROJECT MANAGER (PM) [6]

Here is a brief commentary regarding some of the specific problems that might be facing the project manager.

Project Reports. There are at least two types of reports that need to be prepared for use by the project manager, namely, project cost reports and project schedule reports. These are typically generated by the IT department and not the project itself. Such reports are usually delivered monthly, but when the project nears a significant milestone, that timeline might shrink. If the PM does not get the reports that he or she needs, he is literally "flying blind".

Help from Contracts Department. From time to time, contracts people need to confer with the PM to assure that all contract deliverables are understood and on track. In this case, the contracts people can and should be helping, not beating up on the PM.

Assuring Follow-On Business. The PM must have the follow-on business in his or her sights, all the time. Chase down the opportunities for enhancing the probability of a follow-on contract.

Creeping Requirements. If requirements are creeping, so is the likelihood of failure increasing. The PM must keep the requirements under control and resist any temptation of creep that is embracing by his customer.

Planning All the Time. A project is a dynamic creature. That means that it is changing all the time. Signals for change are provided every time there is a project meeting, especially those that include the customer. Every change needs to be reflected in a change in the project plan. This means, literally, that the PM has pressure to be planning all the time. This will "steal" his time unless he is careful.

Cost out of Control. For more-or-less inexplicable reasons, project costs tend to move out of control. Of course, the PM must make sure that the project stays within budget. A good way to fall a few notches on how the PM is regarded by management is to over-run a contract.

Schedule Slippages. This is another mysterious phenomenon in a project. All is well until and all of a sudden your project is behind schedule. It must be monitored very carefully.

System Not Satisfying All Requirements. Ultimately, the system performance must be measured by the degree to which requirements are satisfied. So we are tracking the two against each other on a continual basis. Alarms must go off for any discrepancy. A typical problem area for IT systems is the response time for the system. It is easy to read and often very difficult to meet. For example, the statement – "the system response time shall not exceed 4 seconds" – can get the PM in a whole lot of trouble.

EARNED VALUE ANALYSIS (EVA)

As the project gets up and running, and time is moving ahead, we are in a position to calculate the following: (1) are we ahead of schedule, (2) are we behind schedule, or (3) are we exactly on schedule? We can then use these rates in time to extrapolate to the expected time to completion. At that point, we have a linear extrapolation (assumption) to what might be the total time for the project. This is the same as the so-called expected total time for the project. It may be compared with the original schedule, and we may say that we are ahead of schedule, behind schedule, or on schedule.

The same basic idea holds for costs – are we overspent, underspent, or exactly on budget, at this time? Using a linear extrapolation, we can then project where we will be with respect to cost when we get to the end of the project. In other words, do we predict that we will be over-run, under-run, or precisely on budget when we get to project completion. Here are the appropriate numerics for the EVA calculations: the three members of the project triumvirate must be able to carry out the estimation on behalf of the overall project.

Selected equations for EVA calculations [7] (p. 155) are given as

The schedule variance $= SV = BCWP - BCWS$

The cost variance $= CV = BCWP - ACWP$

The estimated cost at completion $= ECAC = (ACWP)/BCWP \times BAC$

The estimated time to complete $= ETAC = (BCWS)/BCWP \times TAC$

where BCWP is the budgeted cost of work performed.

BCWS is the budgeted cost of work scheduled.
ACWP is the actual cost of work performed.
BAC is the budget at completion.
TAC is the time at completion.

If we wish to see what the Department of Defense is doing with respect to EVA, they have produced an implementation guide and other documentation that can be readily consulted [8].

AGILE PROJECT MANAGEMENT

This brief overview of agile project management is based upon a description from "**workfront**" [9]. The essential feature of "agile" is the work breakdown into short sprints or iterations. This "chunking" of work is not unfamiliar to software engineers and those adept at constructing work breakdown structures. The sprints, it is claimed, are easier to handle, and usually more rapidly than conventional methods. The **workfront** group has set forth four core values of agile, namely:

1. Prefer focus on individuals and iterations over processes and tools.
2. Prefer operating software over massive documentation.
3. Prefer collaboration with the customer over negotiation over one or more contracts.
4. Prefer responding to more immediate change over following a plan.

Some of the above may well be counter-intuitive in terms of two of the above. First, one might argue that there is a trend toward rather than away from processes. An example of that would be the ISO/IEC 15288 standard. Under "agile", we look more for immediate reaction rather than following a rigid but absolutely process. Second, the fourth item above implies a spontaneity rather than having a plan as a work blueprint.

The **workfront** group has taken the time to define and convey some "12 principles of agile" [6]. The reader can access this information by means of a simple "Google" Search. The bottom line is that there is some belief that "agile" will improve timelines as well as quality, and is a serious candidate to achieve software development improvements.

THE PROJECT MANAGER AND THE MYERS–BRIGGS TYPE INDICATOR (MBTI)

We note the special attention and importance given to communication. This is stressed in this treatise and in consonance with the author's experience. One perspective regarding this important trait is to have the PM better understand his or her project team: simple formula, better understanding, and better communication. In that regard, this short section of the chapter explains a bit about what is known as the Myers–Briggs Type Indicator (MBTI).

The MBTI is a way to measure the "personality" of a person. It is a method that is well-documented and accepted. It's based upon characterizing people with four basic polarities [10], namely:

- Introvert (I) vs. Extrovert (E)
- Intuitive (N) vs. Sensing (S)
- Thinking (T) vs. Feeling (F)
- Judging (J) vs. Perceiving (P).

If we take all the possible combinations of the above, we obtain two to the fourth power or sixteen types. The literature, therefore, is quite extensive regarding each of these sixteen types [4,4].

So here are two reasons to put more attention into how to use the MBTI. The first is to "know thyself", which is likely to make one a better manager. The second is to know your team members, one by one. With the latter, you are more likely to have better communications. Having taken the MBTI some years ago, I do not find it difficult to share my profile with you, the reader. It is INTJ, and apparently, it is shared by only 2.1% of the overall population. There are only two other profiles that are shared by a fewer percent, namely, ENTJ is 1.8%, and INFJ is 1.5%.

Finally, one question of interest here is: is there a preferred MBTI for the role of PM? Or what profile is more likely to lead to successful project management? From the literature, the answer appears to be ESTJ [4]. Does that check out with your idea as to what is important and what is not as important?

ANOTHER PERSONALITY PROFILE

Another profile that I applied to myself and my students had four dimensions, and the overall score added to 40. So the profile was

- Action – 10 points
- People – 10 points
- Process – 10 points
- Idea – 10 points.

For this profile, I elicited an answer from my students as to what might be a profile for an ideal manager. The answer is shown below:

- Action – 11.4 points
- People – 11.6 points
- Process – 9.9 points
- Idea – 7.1 points.

Considering the emphasis these days upon the seminal idea, one might be just a bit surprised at the low score for "idea". But then again, having the best idea and being the best manager are somewhat different skills, as we know.

PROGRAM MANAGEMENT – THE DOD

The Department of Department of Defense (DoD) has made it clear as to what they value with respect to Program Management [11] typically the "boss" of the DoD Project Manager. Their guide, they claim, defined some 80% of what Defense Program Managers need to know to "run an effective and efficient" program. Keep in mind that typically the PM must satisfy the overall condition set by the program manager. Table 6.1 provides a brief overview of the topics they considered important.

TABLE 6.1   Topical guide to program management activities [8]

A. Program Management – The Basics
   A.1 Cost
   A.2 Schedule
   A.3 Performance
   A.4 Risk

B. Financial Management
   B.1 Planning – Cost Estimating
   B.2 Planning – The Color of Money
   B.3 Planning – POM Submissions and Budget Exhibits
   B.4 Congress
   B.5 Executive Phase
   B.6 Driving the Financial Bus

C. The Contract
   C.1 Contract Types

D. Systems Engineering
   D.1 Understanding Systems Engineering
   D.2 Effective Systems Engineering Processes
   D.3 Risk Management
   D.4 Configuration Management
   D.5 Data Management
   D.6 Requirements Management

E. Other Tools and Practices
   E.1 Battle Rhythm
   E.2 EVM – Earned Value
   E.3 External Independent Reviews

F. Acquisition Strategy

G. Acquisition Program Baseline

H. Integrated Master Plan and Schedule

I. Integrity
   I.1 Honesty
   I.2 Completeness
   I.3 Soundness

J. Leadership
   J.1 Definition
   J.2 Vision and Goals
   J.3 Expectations and Responsibility
   J.4 People
   J.5 Personality

(Continued)

TABLE 6.1   (*Continued*) Topical guide to program management activities [8]

K. Collaboration and Compromise
    K.1 Stakeholders
    K.2 The Program Office
    K.3 Good Fences Make Good Neighbors

L. Final Thought – What Does a PM do?
    L.1 Represent the Program
    L.2 Organize for Success
    L.3 Take Care of the People
    L.4 So What Does the PM do Again?

LEADERS FOR SYSTEMS ENGINEERING TEAMS

- There has been an enormous amount of attention paid to the question of leadership, including an earlier book from this author [4]. That treatise cites five attributes of a leader, after due consideration with respect to some 30 sources. Here are the five attributes:
    1. Practical visionary
    2. Inclusive communicator
    3. Positive doer
    4. Renewing facilitator
    5. Principled integrator.

We note that each of the above attributes can be construed as a "double" attribute. This was deliberate, as I recall.

Practical Visionary. This leader has a distinct vision but also takes a practical approach to achieving that vision. One might say that this person is able to easily integrate tactics and strategies.

Inclusive Communicator. This leader maintains communications, making sure that it is inclusive and across-the-board. Recipients of this type of communications appreciate this approach.

Positive Doer. This leader gets things done and is not behind the desk and watching. There may be some difficult tasks to be carried out, but this leader maintains a positive outlook and perspective at all times.

Renewing Facilitator. This leader facilitates by helping others through what might be difficult for them. There is also a sense of renewal by working with this leader so that all feel refreshed during and after tasks are accomplished.

Principled Integrator. This leader demonstrates his or her ability to bring the pieces together; one aspect of the "systems approach". Usually, this is a broader look at the problem so that the pieces "fit" within the suggested solution. At no time, however, does this leader stray from basic personal as well as corporate principles.

If we move forward in time, we come to Norman Augustine's citation of leadership attributes, which are [12]

1. Inspiration
2. Perseverance
3. Courage
4. Selflessness
5. Integrity.

Here, we note that one of the above attributes is the same as one of the attributes of the special systems engineer (i.e., perseverance).

Considering the leadership role with respect to assistance to others, we cite briefly a gurus interpretation of leadership attributes [13], as applied to helping others build their careers. These are

1. Act as facilitator
2. Need to appraise
3. Forecast, as appropriate
4. Advise
5. Enable.

Finally, some quotes from a couple of business gurus (R. Moss Kanter and Peter Drucker) are

Cosmopolitan Leaders of the future must be: [13]

    a. Integrators
    b. Diplomats
    c. Cross-fertilizers
    d. Deep thinkers

Leadership is the lifting of a man's vision to higher heights [14]

The essence of leadership is very simple. It is to help people to perform to their maximum potential to achieve organizational goals or objectives [14]

CLOSING WORDS

This chapter cites a fair number or abilities needed by the systems engineer in order to be successful. It enters into the world of management, often a difficult arena for the systems engineer. If the latter is proficient in most of these skills, then he or she has an improved likelihood of being successful. However, the overall thesis here is that there are still other attributes needed (see Chapter 3). The journey of the extraordinary systems engineer remains a long one. The best in the field leaves a "roadless traveled by".

REFERENCES

1. Kerzner, H., "Project Management – A Systems Approach to Planning, Scheduling and Controlling", 3rd Edition, Van Nostrand Reinhold, 1989.
2. Eisner, H., "Essentials of Project and Systems Engineering Management", 3rd Edition, John Wiley, 2008.
3. Northeastern University, "Breaking into Project Management", 2020.
4. Eisner, H., "Reengineering Yourself and Your Company", Artech House, 2000.
5. See www.pmi.org.uk.
6. ISO/IEC 15288 2002. "Systems Engineering – System Life Cycle Processes," 2015.
7. Kezsbom, D., D. Schilling, and K. Howard, "Dynamic Project Management", John Wiley, 1989.
8. Department of Defense, Earned Value Management Implementation Guide, January 2019.
9. Workfront, "Agile Project Management Methodology", 7 May, 2018, see https:en.wiki.org; workfront
10. Myers, I. B., "Gifts Differing", Davies-Black Publishing, 1980.
11. Cooley, W., and B. Ruhm, "A Guide for DoD Program Managers", Defense Acquisition University Press, December 2014.
12. Augustine, N., "Augustine's Travels", AMACOM, 1998.
13. Kanter, R., "World Class Leaders – The Power of Partnering", from Hesselbein, F., M. Goldsmith and R. Beckhard, "The Leader of the Future", The Drucker Foundation, Future Series, Jossey-Bass, 1996.
14. Cohen, W., "Drucker's Top Tips to Successful Leadership", Lessons from Drucker, 25 June, 2019.

Expert/ESEP

7

INCOSE (International Council on Systems Engineering) has taken the lead with respect to certification of systems engineers. This capability of the systems engineer can be construed to be equivalent to that of the INCOSE ESEP [1]. It is mostly technical in nature and is well-defined in today's current state-of-the-art. Here is some data that contrasts three levels of INCOSE capability [1].

LEVEL	EXPERIENCE REQUIRED	EDUCATION	REFERENCES	KNOWLEDGE
ASEP	No special requirement (Req't)	No special Req't	None required (Req'd)	Written Exam – *Systems Engineering Handbook*
CSEP	Minimum 5 years SE	Qualifying degree	Three references	Written, based on INCOSE
ESEP	Minimum 25 years; minimum 5 years of professional development	Qualifying degree	Three references	Not directly measured

We note the special position occupied by the *INCOSE Systems Engineering Handbook* (Version 4). This is entirely appropriate considering the excellence of that handbook and the fact that it is based upon ISO/IEC 15288 standard. That is where INCOSE is at this point in its history.

SYSTEMS ENGINEERING EXPERIENCE AREAS

Requirements Engineering

Preparing for or managing a business or mission analysis; defining a problem or opportunity space; evaluating alternative solution classes; preparing for stakeholder needs and requirements definition; defining stakeholder needs; developing operational concept and other life cycle concepts; transforming needs into stakeholder requirements; analyzing stakeholder requirements; defining system requirements; analyzing system requirements; and managing system requirements (from INCOSE).

System and decision analysis

Preparing, performing, and managing a system analysis; decision management; preparing for systems engineering decision; analyzing decision information; and making and managing SE decisions (from INCOSE).

Architecture/design development

Preparing for architecture definition; developing architecture viewpoints; developing models and views of candidate architectures; relating architecture to design; assessing candidate architectures; managing the selected architecture; alternatives for obtaining system elements; definition of characteristics and design enablers; and managing a system design (from INCOSE).

Systems Integration

Preparing, performing, and managing system element implementation; identifying, agreeing, and managing system-level interfaces; preparing and performing integration; and managing integration results (from INCOSE).

Verification and Validation

Preparing and performing verification; and managing verification results. Preparing and performing validation managing validation results; preparing and performing system transition; managing results of system transition; and obtaining qualification, certification, and acceptance (from INCOSE).

System Operation and Maintenance

Preparing for operation; managing results of operation; performing and supporting system/product operation; preparing for and performing maintenance; performing logistics support; managing results of maintenance and logistics; preparing for, performing, and finalizing system disposal (from INCOSE).

Technical Planning

Defining an SE project; planning an SE project and its technical management; activating an SE project; identifying and recording tailoring influences and mandated structures; obtaining input from parties affected by the tailoring strategy; making tailoring decisions; and selecting the life cycle processes (from INCOSE).

Technical Monitoring and Control

Planning for SE project assessment and control; assessing SE projects; controlling projects from an SE perspective; preparing for and performing system measurement; preparing for system quality assurance; and performing system product or service evaluations (from INCOSE).

Acquisition and Supply

Acquisition, including preparing for system/element acquisition; advertising the acquisition and selecting the supplier; establishing, maintaining, and monitoring an acquisition agreement; and accepting a product or service from a supplier; and supply, including preparing for supply, responding to a tender; establishing, maintaining, and executing a supply agreement; and delivering and supporting a product or service (from INCOSE).

Information and Configuration Management

Planning configuration management; performing configuration identification; performing configuration change management; performing configuration status accounting; performing configuration evaluation; performing release control; and information management, including preparing for and performing information management (from INCOSE).

Risk and Opportunity Management

Performing technical risk and opportunity management; managing the technical risk profile; and analyzing. Treating and monitoring technical risks and opportunities (from INCOSE).

Life Cycle Process Definition and Management

Establishing life cycle processes including defining and implementing life cycle models; assessing life cycle processes and models; and improving life cycle processes and models (from INCOSE).

Specialty Engineering

Performing professional-level systems engineering activities associated with specialty engineering; manufacturing and produceability analysis; mass properties engineering; reliability, maintainability and availability analysis; resilience engineering; system safety engineering; system security engineering; training needs analysis; usability analysis/human systems integration; and value engineering (from INCOSE).

Organizational Project Enabling Activities

Infrastructure management, including establishing and maintaining the infrastructure; HR management, including identifying and developing SE skills; acquiring and providing SE skills for projects; quality management, including planning and assessment quality management; performing quality management corrective and preventive actions; knowledge management, including planning knowledge management, sharing knowledge and skills throughout the organization; managing knowledge skills and knowledge assets; and project portfolio management at organizational level, including defining and authorizing SE projects, evaluating a portfolio of SE projects, and terminating SE projects (from INCOSE).

DISTRIBUTION OF THE MINIMUM 25/20 YEARS FOR ESEPs

Individual has balance between depth and breadth of SE experience in many, but not all, of SE experience areas: at least 2 years or greater increments in at least 6 of the 14 systems engineering experience areas; all subject to decision of the certification program office.

EDUCATION FOR QUALIFYING DEGREES FOR ESEPs

Technical degree required; minimum of 5 more years of engineering experience with bachelor's degree or minimum of 10+ years of engineering experience with no degree.

For ESEP who is already a CSEP, a total of 25 or 30 years of experience with at least 20 years of SE experience.

For ESEP who is not already a CSEP, this equates to 35 years of experience with at least 25 years of SE experience.

References (personal)

Require at least three references

Exam for ESEPs

Two-hour 120-item multiple choice exam, based upon *INCOSE Systems Engineering Handbook*.

REFERENCE

1. Certification at ESEP Level, "INCOSE (International Council on Systems Engineering)", 7670 Opportunity Road, Suite 220, San Diego, CA, pp. 92111–2222.

Expert/ Domain Knowledge

8

This capability may well be one of the most difficult to obtain, and it refers to expertise regarding the system or domain that the systems engineer is to be called upon to work. An example might be the transition for the information technology (IT) from a "plain vanilla" IT system to one that is fully equipped with crypto capability. The latter is a specialized skill area that is relatively recent. Therefore, the IT engineer may not have run into it in his or her various assignments. It therefore is not part of his or her current capability. It will take some time working in the field to develop what might be judged as true expertise.

In today's world of cyber systems and cyber-technology, these are specialized domains with a lot of off-the-shelf software. It takes quite a while for the systems engineer to become familiar with such software if such is to be part of the system and its architecture. Proficiency is likely to be (a) required and (b) developed only after at least a decade of work in that area. That is the scope of what we are looking for in terms of any specialized domain.

ADVANCED SURFACE MISSILE SYSTEM

This is an example from the author's experience. In my early days as an engineer, I was called upon to participate in a summer study of a technology area that was soon to be instantiated in a real system. The system had the name "ASMS" at that time, which stood for "Advanced Surface Missile System". Although I had a strong education background at the engineering master's level and some background in radar systems, the ASMS technology was mostly new to me. I was certainly not an expert, even in the radar portion of the technology and the system that it might imply.

The overall functional breakdown of a general "ASMS" (unclassified) is shown below:

Functional Breakdown of General Anti-Surface Missile System

    Target Surveillance
    Target Detection
    Target Tracking
    Missile Assignment
    Missile Launch
    Missile Tracking
    Target Hit Assessment.

These seven functional areas look simple enough but embody complex technology. If we dig a bit deeper, we can find missile-related technology that requires many years of expertise in order to be able to contribute. For example, the Missile Defense Agency issued a request for Proposal (Unclassified) to restart a ballistic missile interceptor program that is designed to knock down North Korean missiles that might be attacking us or our allies. This program, called the Next-Generation Interceptor, replaces the Re-designed Kill Vehicle. All of these programs clearly require deep knowledge of both missile technology and kinetics.

RANDOM DOMAINS OF POTENTIAL INTEREST

Scanning my unwanted iPhone messages, we find such entries as follows:

First Scan

- X-37B's Next Mission To Demo Space-Based Solar Power
- The Key to All-Domain Warfare Is "Predictive Analysis"
- NRO, SPACECOM Craft CONOPS For War in Space
- China in Space: Does US Contest or Cooperate?
- RAND Briefs Congress: America Faces Likely NatSec Launch Shortfall
- New NGA Tech Strategy Aims At AI Integration
- Space AQ Council Eyes Helping Space Station Crack CARES Act.

Second Scan

- Esper orders SDA to link C2 Networks For All-Domain Ops
- X-37Bs Next Mission to Demo Space-Based Solar Power
- The key to all0domain warfare is "predictive analysis"
- NRO, SPACECOM Craft CONOPS For War in Space
- RAND briefs congress: America Faces Likely NatSec
- New NGA Tech Strategy Aims At AI Integration.

Third Scan

- Northrop Grumman is Bringing Space Closer to Earth
- India inks $900 million Deal for Sikorsky Sub-Hunting Helos as Tensions with China Spike
- Trump on F-35: We should Make Everything in U.S.
- Roper Targets Commercial AI, Data Analytics for Next ABMS Deals
- MDA: All Domain C2 Key to Countering Hypersonic Missiles.

Fourth Scan

- CQ Brown Brings Pacific Focus; Keen interest in Joint Ops
- Roper Eyes Biotech; New Materials funding; NGAD Gets PEO
- Unanswered Israeli Air Strikes Against Syria Raises S-400 Questions
- Speed, Maneuverability, Survivability and Sustainability Are the Hallmarks of Sikorsky' s X2 and Technology
- GAO Chides DoD for absence of Cybersecurity Requirements
- AFRL Targets Space Ops in New Orbits
- Army tests PrSM Seeker to Hunt Ships and SAMs.

Do we understand even one of these?

Each has its own set of domain knowledge, understood only by those that have been working in the field for many years.

These random scans reveal specialized domains requiring years of involvement before one can be truly productive.

DARPA

For most of us, DARPA operates in several domains of special knowledge. Let's take a brief look at what is on its home page. They reiterate their interest in transformative change rather than incremental change. They do much of their interactions through broad agency announcements. Here is an example of a series of such announcements.

- Atmosphere as a sensor
- Rational integrated design of energetics
- Techniques to characterize the susceptibility of electronics to high-power microwave radiation
- Artificial intelligence exploration opportunity: techniques for machine vision disruption
- Seabed simulation synthesis
- Wearable laser detection and alert system
- Open source wideband software defined acoustic modem
- Ocean of things.

The last-named item certainly gives one food for thought.

NATIONAL AIRSPACE SYSTEM (NAS)

The NAS is a huge system, truly a "system of systems". It is always evolving as new capabilities are added and obsolete sub-systems being retired. A particularly relevant initiative within the FAA is the "Next-Generation System", known as NextGen. This is a look at what the next-generation airspace system will look like. In effect, it's the very preliminary design of such a system.

The FAA claims that NextGen will be a significant improvement in overall capacity, performance, efficiency, safety, environmental factors, and predictability. Its overall functions include

- Navigation
- Communications
- Surveillance
- Automation
- Information management

It is intended to be a modern, resilient and secure system that encompasses innovative and transformative technologies.

OTHER POTENTIAL DOMAIN AREAS

There are literally hundreds of specialized domain areas in which the systems engineer might work. We obtain some idea as to domains by enumerating the typical departments at your local university. That is what was done to formulate the following table:

POTENTIAL DOMAINS FOR SUPPORT FROM THE SYSTEMS ENGINEER [1,2] (*)

Classical Acting	Accountancy (*)	American Studies (*)
Anatomy	Anthropology (*)	Applied Science (*)
Arabic	Art History	Art Therapy (*)
Astronomy (*)	Biochemistry (*)	Biological Sciences (*)
Biomedical Sciences	Biostatistics (*)	Business Administration (*)
Chemistry (*)	Chinese	Civil Engineering (*)
Classical Studies	Communication	Computer Science (*)
Counseling	Decision Sciences (*)	East Asian Languages
East Asian Literature	Economics (*)	Educational Leadership
Electrical Engineering (*)	Computer Engineering (*)	Emergency Health Services
Engineering Management (*)	Systems Engineering (*)	English
Environmental Policy (*)	Epidemiology (*)	Exercise/sport activity
Exercise Science	Film Studies	Finance (*)
Fine Arts	Forensic Sciences	French
Fine Arts	Forensic Sciences	French
Geography	Geographical Sciences	German
Greek	Health Care/Sciences	Hebrew
History	Hominid Paleobiology	Human Development
Organizational Learning	Humanities	Information Systems (*)
Technology Management (*)	Interior Design	International Affairs
Italian	Japanese	Korean
Landscape Design	Latin	Law

(Continued)

POTENTIAL DOMAINS FOR SUPPORT FROM THE SYSTEMS ENGINEER [1,2] (*)

Law Firm Management	Linguistics	Management (*)
Marketing	Mathematics (*)	Mechanical Engineering (*)
Aerospace Engineering (*)	Microbiology	Immunology
Molecular Biology	Molecular Medicine	Museum Studies
Music	Naval Science	Organizational Sciences
Paralegal Studies	Pathology	Peace Studies
Persian	Pharmacology	Philosophy
Physics	Physiology	Political Management
Political Science/ Psychology	Portuguese	Psychology
Public Administration	Public Health	Public Leadership
Public Policy	Public Relations	Publishing
Religion	Romance Literatures	Security and Safety
Slavic	Sociology	Spanish
Special Education	Speech and Hearing	Statistics
Strategic Management	Teacher Education	Theater and Dance
Tourism Studies	Turkish	University Writing
Vietnamese	Women's Leadership	Women's Studies
Yiddish		

(*) More likely to have support from systems engineers.

REFERENCES

1. See https://en.wikipedia.org./wiki/Next_Generation_Air_Transportation_System.
2. George Washington University Class Bulletin.

Perseverer

9

This chapter completes the presentation of "success" skills that the super-systems engineer is asked to have. This skill has to do with persevering, no matter the barriers and no matter the hurdles. There are many very strong systems engineers that can tackle very difficult problems; not every systems engineer will be able to put forth a truly persevering effort. Another word that we have run into in this regard is "grit" [1]. It is a word and concept that has many dimensions.

A very simple example of perseverance was set forth by one of our great inventors. The quote was "Invention is 1% inspiration and 99% perspiration". He certainly knew that territory – he had more than 1000 patents that accrued from his work. One would imagine that he had many moments when he wanted to slow down or stop, but he didn't. He was, by all accounts, our number one "perseverer". An elaboration of what Edison said has appeared in the literature, and is something like that there are three essentials to being successful: hard work, stick-to-it-tiveness, and common sense. It's hard to misunderstand what it is that guided Edison his whole life, and found most important.

Another example of just plain perseverance is that exhibited by Robert Pirsig in trying to get his book published [2]. According to several accounts, Pirsig went to over 100 publishers before he found one that would agree to publish his book. He persevered, and it paid off. It usually does, given a level of quality of the original product (or service). Let us explore some of the dimensions of perseverance or grit.

WATCH THAT EYEPATCH

So the first "True Grit" film came out in 1969 with John Wayne as Rooster Cogburn, Kim Darby as the heroine who hires Cogburn to find her father's killer, and Robert Duvall as the killer. The second film, by the Coen brothers, came out in 2010 with Jeff Bridges as Rooster Cogburn, Hailee Steinfeld as the heroine, and Josh Brolin as the killer/outlaw. Rooster, in each case, sports a fancy and notable eyepatch. Both films were actually called "True Grit" [3,4]. And who was it that had the true grit? As best I could tell, both Rooster and the girl had oodles and oodles of it. And we were rooting for them from pillar to post. Classic heroes and heroines, and what could be more fun than seeing them fight against all odds and achieve the success they sought. And who could fail to recall the scene, in both movies, when Rooster had his showdown with the bad guys? He grabbed the reins and charged, shooting with what seemed like an arsenal of guns. "You will fall to my onslaught" he was saying with his actions, and his just plain pure grit.

GRIT, THE BOOK

It was not too many years after the second True Grit movie, a book came out that addressed the matter of "Grit", true or otherwise [1]. The book is sub-titled "the Power of Passion and Perseverance" – and so it is. It was a block-buster success, written by Angela Duckworth, a PhD MacArthur Fellow.

If we had to sum up the message from Dr. Duckworth, it might well be subsumed in two "equations":

$$\text{Talent} \times \text{Effort} = \text{Skill}$$

$$\text{Skill} \times \text{Effort} = \text{Achievement}$$

Looks simple, and intuitively appealing. You've got to put in the effort if you want to be successful. And how many of us want to be successful? The answer? All of us. And how many of us are willing to put in the effort? Answer? Not all of us.

So we are using "grit" and "perseverance" interchangeably. What does our friend, the "New World Dictionary", say about that?

Definition – Grit = Stubborn courage, brave perseverance, pluck

That's good enough for this author. I will use these words interchangeably.

Dr. Duckworth believes that perseverance is measured by one's ability to overcome setbacks, put in the necessary hard work, and complete tasks that you've started. That's a mouthful, and each part may be elaborated upon.

Overcoming Setbacks. It depends a lot upon the strength of the setback. Some setbacks are just too tall, too imposing. But you get the idea – you find a way to overcome and that may be dodging and walking around. It may be getting some help. It may be a story all to itself.

Put In the Hard Work. That's pretty broad, but we know that many just plain cave when the going gets tough. Just too much effort required; don't have it in you. Never learned how to do that. What size incentive will be enough to get you to put in that hard work?

Complete the Tasks You've Started. How many of us have unfinished business out there? How many of us have our desktops cluttered with papers that are the beginnings of many projects and new ideas? How many of us will never finish because we deep down don't want to finish? This, too, is a bit more complicated than it looks.

Now – how does talent fit into the equation? Well, you're glad to have the talent, but it's by no means the whole story. What about the three factors listed above? We're happy to have the talent, but it's not enough. You need more than that. Remember that "effort" counts twice:

$$Skill = Effort \times Talent$$

$$Achievement = Effort \times Skill$$

So – Dr. Duckworth comes to this conclusion: the more effort you apply, the more your skill increases, and the more you are likely to achieve. Grit is a predictor of success, and two people of roughly the same talent will see the one with more grit succeed vs. the other with less. Quite a differentiator, isn't it?

PRACTICE AND EFFORT

Due to the central position taken by "effort", it would seem that a few more words about it are appropriate. For purposes of this exposition, we will take "practice" and "effort" to be just about the same. However, we need to be somewhat careful about accepting the former without reservation.

First, we have the observation that this man is running up and down 57th Street and 7th Avenue in Manhattan when he finally stops a more elderly man on his side of the street.

"Please tell me, he asks. How can I get to Carnegie Hall?"

The older man pauses before he answers.

"Practice, my boy…practice", he repeats.

A great anecdote for most of us. But here's some more about "practicing".

My boss, a vice president at my company, tells this story about his boss, the company's president. The president was a very determined man with what would appear to have considerable perseverance. That led him, without coaching, to lots and lot of practice with playing golf which he wanted to improve. In particular, he wanted to get better at the long drive from the tee. So he bought several buckets of balls, and teed off with them for a couple of hours. When he was done, he backed up, breathing hard, with hands that were raw and almost bleeding. He found that very satisfying, my boss reports, but my boss saw some of it first hand and was not able to see any improvement. In fact, he observed shots that seemed to be getting worse with time. What are we to say about this?

My boss hit the nail on the head. Practice is great, but one must be practicing "the right" way. If your form and shot are terrible, then all you'll be doing is reinforcing the wrong way to tee off. And after many buckets of balls, you'll have that down pat. You'll really know how to hit the ball the wrong way.

So there's a footnote to this matter of practicing – you'd best be practicing the right way. And that's why we take lessons from a pro, to show us the right way. And then, we copy what they tell us, without going off on our own with new moves.

So another footnote that applies to the life of this author – my piano practice at age eight was less than steady, and definitely not what my teacher was trying to teach me. Do you have the same story to tell? What about this business of practice?

Dr. Duckworth has not missed this point. She defines an activity known as "deliberate practice" which is basically practicing the right way, with deliberation. If you can figure this out, you're on the right track. If not, you're still hitting those buckets of golf balls. Duckworth uses Ben Franklin as a positive example to improve his writing. Franklin poured over his text, found the faults, and corrected each and every one until it was as perfect as humanly possible. In this case, Franklin was his own critic. Not a bad critic to have if you want to improve.

So the sequence appears to be: determined practice → deliberate practice → improved effort → improved skill → improved effort → improved achievement.

We each have our way, but there are some basics, such as the above sequence. And Duckworth claims that even Kevin Durant has his own way:

> "I probably spend 70% of my time by myself, working on my game, just trying to fine-tune every single piece of my game". I think he understands, for himself, the meaning of "deliberate". What would be your definition of "deliberate"?

BEZOS AND SHORTZ

It was easy to see how it was that Duckworth hammered out a best seller. She had a great story to tell, and made sure that she showed real world example after example. Two of those examples were with Jeff Bezos and Will Shortz. Bezos showed great passion for home-made electro-mechanical projects, and Shortz immersed himself with everything that had to do with crossword and other puzzles. So the passions were there, and the talent was there, and the effort was there. What else was needed?

That "what else" was perseverance and grit. Here's what that eventually resulted in.

First, Bezos graduated from Princeton at the top of his class and took an early job in the banking industry. After a few years of getting grounded in that business area as well as in the technology, he opened up an on-line bookstore which he called Amazon. This was not received with enthusiasm. How could you abandon the banking business (where there was money to be made) in favor of a bookstore (where it's not clear there's money to be made). Nonetheless, he persisted and went for an IPO (initial public offering) 3 years after its creation. He was not "making money" (read "profits"), but he kept going and going and going. He took the critical step of moving Amazon into products beyond books, and featured low-end prices as well as extraordinary delivery service. Many said that such a strategy would bring bankruptcy rather than profitability. But what did Bezos wind up doing? He seemed to double-down over and over until he became the wealthiest person in the United States. He didn't follow the advice of quarter by quarter profits, but he literally wrote the book on being the best on price and service. Sound familiar? And as we speak, we're still following the adventures of Jeff Bezos as he leads with his unconventional approach, breaking new and old rules with his ideas and perseverance.

Will Shortz is the *New York Times* crossword puzzle expert, so that when you see their puzzle, it's likely it was constructed by Will. But beyond that, I also consider him to be a supreme puzzle-master. My evidence of this exalted position is that just about every Sunday morning I have encountered Will on FM radio, putting a word puzzle out there and hosting a show dealing with that puzzle. So Will is the real thing – making and solving puzzles is his main game. No doubt that he's an expert, reaching the top of his field. That probably also means that he has satisfied that conventional definition of an expert – put in 10,000 hours of practice over a ten-year period. So since I meet with Will every Sunday morning, I'm happy to recommend him even though I don't do the *New York Times* crossword puzzle. And by the way,

Duckworth's book contains Shortz's suggestion as to how to do the *New York Times* crossword puzzle:

1. Begin with the answers you're surest of.
2. Don't be afraid to guess.
3. Erase an answer that doesn't seem to work.
4. So work in pencil.

FINAL OBSERVATIONS FROM DR. DUCKWORTH

This extraordinary book has many words of wisdom that are worth noting before we move on to the next subject.

Kaizen. Dr. Duckworth brings us back to the time when we were struggling with the "quality" issue in the United States. She reiterates the need for Kaizen – continuous improvement. Not new, but well worth thinking about again.

Developing Skills. To say it again, so that it's not forgotten. If you want to become an expert in any given field, it will take at least 10,000 hours of deliberate practice to have any chance of achieving.

Purpose. Keep in mind that you are working toward one or more goals, with a purpose. It is likely that that purpose is to help others, to pay dividends for others. If that's not your purpose, write down what your purpose is and stare at it for a while. If you are okay with the purpose you have chosen, that should be enough of a motivation to continue on your path.

Growth Mindset. Behind just about all the suggestions is the notion that you, and your enterprise, have implicitly accepted a "growth mindset". If that is not true, what is true? What is your mindset?

Grit Culture. Some enterprises have deeply accepted the notion that they are committed to a "grit culture". Duckworth spends some time referring to this on the part of the football team, the Seattle Seahawks. Examine this question. Are you and your organization so committed? Does this whole issue need to be re-examined and explored? Remember – a grit culture can be grown. Is that a goal you've overlooked?

Dimensions of Character. It is defined in terms of strengths of will, heart, and mind.

Other Success Factors. Duckworth answers the question – is grit the only psychological factor that determines success? Answer: Not at all. Consider emotional intelligence, physical talent, intelligence, conscientiousness, self-control, imagination, luck, and opportunity. The list goes on…and on.

EINSTEIN'S PERSEVERANCE

Back in Chapter 2, we made a short reference to Einstein's perseverance and called the eclipse of 1919 a turning point in his life. Actually, there were several turning points, and his perseverance was tested in each one of them [5]. He was supposed to be physics' shooting star, but he hit some obstacles that set him back and required patience and perseverance.

His early school days were beset by teachers that did not appreciate him, to say the least. A notable point, in 1898, came when he applied for entry into the Munich Technical Institute. He was rejected so you can imagine where he was at that time. He also was having not a wonderful time with members of the opposite sex, and by his own admission, he was a mess. A 2016 biographer took note of who he was and what he needed [6]. The article blazed with

Lesson from Einstein: Genius Needs Perseverance

And, indeed as it turned out, he needed it and he embraced it. If he rode that wave, can you also do so?

RICKOVER'S PERSEVERANCE

Can you imagine what amount of perseverance a nice, short, frail, Jewish boy needed in order to

    a. Make it to Admiral in the Navy
    b. Lead the critically important nuclear submarine force, one-third of
       the U.S. triad

Rickover clearly understood that he had an obligation and a commitment to the Navy and his country. He was uniquely placed to carry out this three-decade assignment. Here are some stories about Rickover that are worth noting in the context of his solid perseverance.

OPPENHEIMER'S PERSEVERANCE

As we know, J. Robert Oppenheimer was selected to run the Manhattan Project, which was our building of the atomic bomb. The project was situated at Los Alamos Laboratory, and all witnessed the first detonation on July 16, 1945. Oppenheimer was severely affected by this power and the damage that it did and was capable of doing. He turned to Hinduism for solace, including the Hindu scripture of the *Bhagavad Gita*. The phrase that he found helpful, for whatever reason, was

> Now I am become death, the destroyer of worlds.

But nonetheless, he continued on…he persisted. He showed perseverance in a time of great personal stress.

PERSEVERANCE AND RESILIENCE [6]

In an article back in 2017, Jay Wren declared that "persistence and resilience are the power tools of success". Beyond the declaration, he gave us four examples.

Robert Pirsig. He wrote "Zen and the Art of Motorcycle Maintenance", and then set about to have it published. It was rejected by some 121 publishers and then finally accepted by the 122nd. How's that for persistence? Of course in retrospect, it was all just a fun story of extraordinary patience. But how many of us would get to the end game, as did Pirsig?

J. K. Rowling. This, of course, is the well-known story of the Harry Potter series. Lots of rejections, but continue writing and continuing to believe. Here again, how many of us would be able to demonstrate the needed perseverance as well as resilience? And let us not forget the talent as well as the imagination.

Oprah Winfrey. She was born into poverty and picked her head up long enough to win the Miss Black America beauty context. Still in high school, she got a part-time job as a radio news anchor. Two years later she became a television news anchor. This put her in front of larger and larger crowds that she charmed with her intelligence and wit. Who could resist her, at that point? And who could resist her now?

Steven Spielberg. He tried to get into the University of Southern California but was twice rejected. So he turned around (some years later) and found his way to a degree at the California State University Long Beach. But all that was after recognition with Oscars, Emmys, golden gloves, and some honorary doctorate degrees. Who knew that Spielberg had it in him? Answer. He did, and he figured out how to do all of it without the degree, the piece of paper. We have to remember to ask Steven how he did that, and what he would recommend today.

These are astonishing stories, and we recognize the hero in each and every story. So it's possible to rise above it all, as they demonstrated. All kinds of blood, sweat, and tears. All kinds of persistence, resilience, and grit.

REFERENCES

1. Duckworth, A., "Grit", Scribner, 2016.
2. Pirsig, R., "Zen and the Art of Motorcycle Maintenance", HarperTorch, 2006.
3. Hathaway, H., "True Grit", Paramount Pictures, 1969.
4. Coen, J. & Coen, E., "True Grit", Paramount Pictures, 2010.
5. Dorminey, B., "Lesson from Einstein: Genius Needs Perseverance", *Science*, October 2016.
6. Wren, J., "Achieving Success Through Persistence and Resilience", see www.jaywren.com, 8 May, 2017.

Recapitulation

10

SYNTHESIZER

First step lies in system architecting.

Initial guidance with respect to architecting is DoDAF.

New and recommended approach to architecting is EAM, Eisner's Architecting Method.

Four steps of EAM are

1. Functional decomposition
2. Synthesis
3. Analysis
4. Cost-effectiveness evaluation.

Developing alternative architectures are part of synthesis.

The synthesizer is expert at systems integration.

Specific suggestions for integration.

Further examples of synthesis.

- A. D. Hall
- Aviation Advisory Commission
- Air Defense System
- Synthesis step as part of systems architecting

Analysis of alternatives (AoA).

Cost-effectiveness analysis.

LISTENER

The Challenger accident, with Feynman listening very carefully
    Corporate listening
    Corporate culture of listening
    Hewlett-Packard's culture of listening
    Active listening
    The art of listening
    One woman's view of listening
    One man's view of listening
    One Dean's view of listening
    Listening as engineering communication
    Communities of practice (COP)
    Does not listening lead to Groupthink?
    One-on-one not listening.

CURIOUS/SYSTEMS THINKER

    Quotes with respect to curiosity
    da Vinci had an "insatiable curiosity"
    Key systems questions that flow from curiosity

        General systems theory
        Ubiquitous MBSE
        Software and systems architecture
        National aviation systems model (NAS)
        Air defense model
        International negotiation model.

    Key variables for a NAS model
    The Lemelson Center
    Strategic plan of Center
    A consequence of curiosity – thinking
    Diversity of thinking approaches

        Inductive thinking
        Deductive thinking

Out-of-the-box thinking
Reductionist thinking
Systems thinking
Design thinking
Disruptive thinking
Lateral thinking
Critical thinking
Hybrid thinking
Breakthrough thinking
Fast and slow thinking
Visual thinking.

Kahneman and Tversky propositions

Regression to the mean
Loss avoidance
Representativeness and availability.

Groupthink
The systems thinker
The five disciplines – Peter Senge
The systems approach.

- Establish and follow a systematic process
- Assure interoperability and harmonious system operation
- Be dedicated to the consideration of alternatives
- Use iterations to refine and converge
- Create a robust and slow-die system
- Satisfy all agreed-upon requirements
- Provide a cost-effective solution
- Assure the systems' sustainability
- Utilize advanced technology, at appropriate levels of risk
- Employ systems thinking.

MANAGER-LEADER

- Manager as project leader
- Core competencies of project manager
- UK perspective on project manager competencies
- UK Project Management Institute

- Problems facing the Project Manager
- Earned Value Analysis
- Agile Project Management
- The Myers–Briggs Type Indicator
- Another personality profile
- Program management – DoD
- Leaders for systems engineering teams
- Augustine's leadership attributes.

EXPERT – ESEP

INCOSE and the ESEP

Experience areas

- Requirements engineering
- System and decision analysis
- Architecture/design development
- Systems integration
- Verification and validation
- System operation and maintenance
- Technical planning
- Technical monitoring and control
- Acquisition and supply
- Information and configuration management
- Risk and opportunity management
- Specialty engineering.

EXPERT – DOMAIN KNOWLEDGE

- Functional breakdown
- Anti-surface missile system
- Random domains of potential interest
- DARPA
- National Airspace System
- Other potential domain areas.

PERSEVERER

- Watch that eyepatch
- *Grit*, the book
- Final observations from Dr. Duckworth
- Perseverance and resilience
- Bezos and Shortz.

SUMMARY

Here's the list of success factors for the systems engineer, one more time.

1. Synthesizer
2. Listener
3. Curious/Systems thinker
4. Manager/Leader
5. Expert – ESEP
6. Expert – domain knowledge
7. Perseverer.

Appendix A – INCOSE Fellow Inputs

This author queried the INCOSE Fellows [1] on the matter of what it is that makes a successful systems engineer. The language of this query was

> Dear fellow Fellows,
> It is my intention to construct a short book on the attributes of the successful systems engineer. So – I ask for your opinion: what do you think it takes to be a successful working systems engineer? To sharpen the question: please identify 7 fundamental attributes of the successful systems engineer (not in any order). Please send your list to me at your convenience and as soon as possible.
>   Two notes:
>
> 1. "Successful" is how you would define it.
> 2. "If you send me your opinion, that will constitute your permission to include your results, without attribution".

Inputs were received, but not enough of them to constitute a probability (statistical) sample. Thus, this appendix provides a brief discussion of some of the inputs from each Fellow that did respond. This approach is literate rather than numerate.

WOLT FABRYCKY

1. Has endured the rigor and discipline of an academic degree in engineering
2. Has engaged in interdisciplinary intellectual growth and understanding
3. Thinks about the system life cycle end before beginning per da Vinci
4. Focuses on what the system does before what the system is

5. Becomes and applies PE registration intent beyond the physical
6. Employs the design dependent parameter paradigm over the life cycle
7. Engages in computer model based systems engineering.

ERIC HONOUR

Identified as essential for success:

1. Product skills and knowledge (in the domain being worked)
2. Understanding and translating stakeholder needs into technical
3. Technical analysis to validate the technical understanding
4. Creative architectural design to implement the technical understanding
5. Team leadership skills
6. Negotiation skills, with the ability to influence the organization
7. Project management skills, ability to plan, organize, and track project tasks
8. SE process knowledge, to know what will be useful next, – tempered by knowledge of how to minimize process overhead
9. Toolkit of generally useful skills such as
   a. Diagramming and modeling (today this is often called MBSE
   b. Technical/cost/schedule tradeoffs
   c. Risk and opportunity management
   d. Configuration management, including requirements management.

SCOTT JACKSON

Two most outstanding attributes deduced from experience:

a. Knows what the SOI is. In most cases, it is the whole system itself (e.g., an aircraft and interfacing systems).
b. Successful SEs do not recognize "rankism" as a valid idea; that is, a person is right just due to his or her organizational rank.

c. Unsuccessful SEs think that SE can be performed within a single group, e.g., avionics. These people are just wrong. This would be sub-systems engineering.

d. Successful SEs recognize that the customer is not always right. The successful SE knows that he or she needs to explain the consequences of their decisions to the customer.

e. Successful SEs know that decisions can be distorted by prior beliefs and emotion and that independent review of decisions is needed.

f. Successful SEs know that risk is an inherent property of all projects and that it is not shameful to address these risks.

g. Successful SEs know that very few executives are capable of designing a system and that "deferring to expertise" is the driving principle.

JOSEPH KASSER

Sent copy of Paper "A Maturity Model for the Competency of Systems Engineers", by Joe Kasser and Moti Frank, Proceedings of the 20th International Symposium of the INCOSE, Chicago, Il, 2010.

KNOWLEDGE, SKILLS, AND ABILITIES (KSA)

The Vertical Dimension

1. **Knowledge** of systems engineering and the application domain in which the systems engineering is being applied.

2. **Cognitive Characteristics**, namely, the ability to think, identify, and tackle problems by solving, resolving, dissolving, or absolving the problems (note Ackoff).

3. **Individual Traits**, namely, the ability to communicate with, work with, lead, and influence other people. Also include personal relationships, team playing, negotiating, self-learning, establishing trust, and managing.

   a. Systems thinking
   b. Big picture perspective
   c. Operational perspective

  d. Functional perspective
  e. Structural perspective
  f. Generic perspective
  g. Continuum perspective
  h. Temporal perspective
  i. Quantitative perspective
  j. Scientific perspective.

Critical Thinking (from Wolcott)

  a. Strategic re-visioner
  b. Pragmatic performer
  c. Perpetual analyzer
  d. Biased jumper
  e. Confused fact finder.

Cognitive Characteristics

1. Understanding the whole system and seeing the big picture
2. Understanding the interconnections between elements and subsystems
3. Understanding systems without getting bogged down in details
4. Having a tolerance for ambiguity and uncertainty
5. Understanding the implications of a proposed change
6. Understanding a new system/concept immediately upon presentation
7. Understanding analogies/parallelism between systems
8. Understanding and exploring synergy
9. Thinking creatively.

MARK MAIER

My personal list of attributes to be a successful systems engineer:

1. **Problem Domain Knowledge**. Significant knowledge of an important problem domain where systems are developed to address said problems.

2. **Solution Domain Knowledge**. Significant knowledge of implementation technologies relevant to one's problem domain. This might be aircraft, electronics, software, or whatever.

3. **A Holistic Perspective**. You can cross back and forth between problem and solution domain without being fixate on either one. You understand the nature of the system as a whole.

4. **Comfortable Working in Abstractions**. Comfortable working with concepts like function and capability abstracted away from a specific implementation, but retaining the quantitative aspects even while abstracted.

5. **Facility with the "Science" Aspects of Modeling**. Ability to use relevant modeling techniques and the discipline to compose correct and complete models.

6. **Facility with the "Art" Aspects of Modeling**. Ability to determine the most effective level of abstraction in a model, being able to identify the most relevant and essential features and capture them while filtering out details that can be (and should be) suppressed.

7. **Embracing Continuous Learning**. The SE needs the ability and the willingness to keep learning and changing as the demands of the environment change.

On the research and empirical side, I think there are some interesting works:

Davidz, Heidi I., and Deborah J. Nightingale, "Enabling systems thinking to accelerate the development of senior systems engineers", *Systems Engineering* 11.1 (2008): 1–14.

Valerdi, Ricardo, and Heidi L. Davidz, "Empirical research in systems engineering: challenges and opportunities of a new frontier", *Systems Engineering* 12.2 (2009): 169–181.

Wood, Danielle, and Annalisa Weigel, "Charting the evolution of satellite programs in developing countries – The Space Technology Ladder", *Space Policy* 28.1 (2012): 15–24.

Frank, Moti, "Knowledge, abilities, cognitive characteristics and behavioral competences of engineers with high capacity for engineering systems thinking (CEST)", *Systems Engineering* 9.2 (2006): 9–103.

GREGORY PARNELL

List of SE attributes

1. ABET-accredited degree and ability to learn and understand new technologies.
2. Focus on system value: Strive to identify and create value for customers and stakeholders.
3. Have good soft skills: Listen to team members and can succinctly communicate with senior leaders and stakeholders.
4. Use systems thinking: Always think about system behavior and take a systems view.
5. Are creative: Seek novel ways to view and solve problems.
6. Understands the mathematical foundation of each SE tool and techniques.
7. Ability to use probability to understand and model uncertainty and risk.
8. Can tailor SE tools and techniques. Know the role of SE tools and techniques and can tailor them to each system.
9. Lead teams: Have leadership skills to lead diverse group members to achieve common objectives.

SARAH SHEARD

Basically,

1. Big picture thinker
2. Ability to focus on who the stakeholder is and what they want (not just what the builders want to build).
3. Ability to focus on the service the thing to be built will provide, not just what it'll look like/feel like/do.
4. Ability to "shoot the breeze" technically.
5. ADD…seriously, to be able to shift focus from technical to managerial to people and back in quick succession.
6. Can "do the math" and can also write/speak coherently.
7. Can follow scenarios in head and look for trouble…will this work if this, this and this happen? What if cybersecurity scenario X happens? What if divide by zero? What if? What if what if?

8. Can speak lots of different technical languages...wait a minute, the thermal guys are planning on xxx but if that's the case, is that going to interfere with your plans to ...?
9. Optimistic and encouraging of people but pessimistic and suspicious of data.

HILLARY SILLITTO

Offers top three (on a non-exclusive) basis:

1. Tolerance for ambiguity
2. Ability to frame and ask the right questions in a domain outside their core expertise
3. Follow the evidence.

CHARLES WASSON

Preamble

Over the past 60+ years, SE has implicitly "branded" itself as an engineering management (EM) and processes discipline to solve early 1950s management engineering workflow issues (Wasson 2018). SE evolved into a 90% management and 10% technical emphasis via its standards, handbooks, textbooks, courses, and certifications, at the expense of SE technical competency, which was ignored. Despite SEs EM "talking points", technical projects for large complex systems continue to exhibit technical, cost, and schedule performance issues traceable to this transformation. As a result, engineering discipline and others

a. View SE as a "mile-wide and an inch deep"
b. View SEs as project coordinators and communicators, not engineers
c. Ask:
   1. Is SE really an engineering discipline or a profession?
   2. Where is the engineering technical discipline and competency?
d. Some suggest that SE should be moved to academic business management programs.

The reality is: bonafide "systems engineering", like other engineering disciplines, has a balanced 90% technical – 10% management mix based on its engineering foundation of concepts, principles, and practices.

Recognize that the attributes listed below apply to both the SE EM and engineering discipline contexts. An uninformed manager can unwittingly employ someone as an SE with only EM dominant skills and limited SE technical competency, be disappointed in their technical performance results, and their own decision-making. Unfortunately, SE, as an engineering discipline, is often erroneously blamed for these poor staffing selection decisions.

- Wasson, Charles S., 2018, SE Management is Not SE Core Competency; it's time to shift this outdated, 60+-year-old paradigm, *Proceedings of the 28th Anniversary of the INCOSE International Symposium (2018)*, Washington, DC (US).

Prerequisite Knowledge, Skills, and Abilities (KSAs) for an SE

a. **Passion**. A self-motivated interest in becoming an SE with the recognition and understanding that an SE title is "earned" via success and respect of their peers over many years.

b. **Education**. A competent understanding of an accredited engineering discipline, its taxonomy, and physics; their concepts, principles, and practices.

c. **Diverse General Understanding**. A competent understanding of an engineering discipline, specialty engineering disciplines, and physics; technologies; tools; multi-level specification requirements development, analysis, flow down, and traceability; modeling and simulation (M & S); bonafide SE as an enabler for MBSE; etc. (A requirements engineer is a "requirements engineer", not an SE.)

d. **Experiential Knowledge**. 20+ years of experience in several small-to-large, end-to-end projects for external customers.

e. **Interpersonal Skills**
   e1. **Communicator**. Ability to communicate effectively with customers, management, vendors at all levels, etc. using verbal, written, and presentation skills.
   e2. **Leadership**. Ability to communicate a development solution vision effectively; serve as a team educator, trainer and mentor; develop team synergy; maintain team convergence and focus on the key issues and completion; establish team priorities; conduct constructive critiques; and resolve conflicts.

f. **Systems Thinker**. Ability to apply Systems Thinking for bounding and specifying a system's context; observing existing system behavior and performance; and understanding the user's vision, context, operating environment, issues, and challenges.

Requisite KSAs for Designation for SE Job Title

a. **Strategic and Tactical Planner**. Ability to apply systems thinking to strategic and tactical planning to determine the resources – budget, schedule, personnel, KSAs, tools, processes, and methods – required to develop a system design solution, assess risk for shortfalls, determine risk and risk mitigation, and apply corrective actions for incrementally managing successful completion of the project.

b. **Technical Project Manager**. Ability to serve as a liaison between the PM and engineering teams to accomplish technical, cost, schedule, technology, and risk programmatics efficiently and effectively.

c. **Leadership**. Ability to keep the development team focused, maintain development system integrity, and "keep the project sold" professionally and ethically based on facts, not spin.

d. **User Advocate**. Serve as a user's advocate to ensure the total System Design Solution focuses on the integration of the users as an integral performance element of the system rather than focusing on the system as a "box".

e. **Stakeholder Identification and Requirements**. Ability to identify key stakeholders and their operational needs; and translate those needs into system capabilities and specification requirements.

f. **System Integrator**. Ability to seamlessly integrate multiple disciplines into a development and design solution, mature the solution, and resolve technical and organizational issues beginning on Day #1.

g. **System Architect**. Ability to translate a set of system or lower level specification requirements into a set of viable candidate architectures for evaluation via an Analysis of Alternatives (AoA) tradeoffs based on user-values factors for selection of an optimal architecture that complies with and can be *verified* as meeting the specification requirements and *validated* by the user in their actual operating environment.

h. **Baseline Manager**. Ability to maintain *informal* discipline and control of the evolving system design solution prior to *formal* baselining and control from that point forward.

i. **Development Environment**. Ability to formulate and mature an Engineering Development Environment (EDE) consisting of an integrated set of compatible tools, methods, and processes that meet the needs of the development teams within a constrained set of resources.

j. **System Integration and Test planner**. Ability to plan, assess, and evaluate test plans for verifying the greatest number of specification requirements with the least number of tests; collecting facts concerning incidents, hazards, and events; isolating probable causes and effects; analyzing system performance.

k. **System Thinker**. Ability to analyze, visualize, reason, prioritize, and strategize "cause-and-effect" relationships in

l. Identifying the sources of requirements and design deficiencies and flaws, anomalies, and defects.

REFERENCE

1. There are approximately 80 INCOSE Fellows as of the writing of this book.

Appendix B – Across the Board Articles

SPEC INNOVATIONS ARTICLE [1]

This input is from a professional systems engineering company rather than a single person.

1. **Patience and Perseverance**. This input correlates 100% with one of the suggested qualities, namely, perseverance. It suggests that "patience" be coupled with perseverance and thus somewhat broadens the overall input.
2. **Know When You Are Done**. This input does not correlate well with the suggested qualities. Of course it is important to know when to stop, as it is with most endeavors.
3. **Have an Analytical Brain**. This input correlates well with the suggested qualities, in particular, "systems thinking". We accept the fact that they are not exactly the same and also the requirement that the systems engineer be gifted in terms of analytical thinking.
4. **Knowledge of Systems Engineering Software Tools**. Tools that are applicable to the tasks and activities of systems engineering are explored in the author's book [2].
5. **Strong Organizational Skills**. This connects to capabilities as a manager.
6. **Ability to See the Small Picture**. This tends to focus on details such as the Work Breakdown Structure (WBS) and other project management skills.
7. **Ability to See the Big Picture**. This pertains to systems thinking, and broadening of one's perspective regarding any given problem.
8. **Well-Rounded Background**. This is interpreted as capable and comfortable with both right and left brain thinking; tendency to be an excellent reader.
9. **Communication Skills**. Strong managers almost always have this capability with members of the team as well as personnel outside the immediate team.

10. **Ability to Lead, Follow, and Work Well in a Team**. This is their definition of leadership.

NEW ENGINEER ARTICLE [3]

This second article deals approximately with the same subject as the first article, namely, "The ten characteristics of a successful engineer". It considers the engineer in general, in distinction to the systems engineer, and it connects well with the attribute of manager/leader. That difference in scope should show up in comparing sets of attributes. The set of ten characteristics for this article is provided as follows:

1. **A Team Player**. This engineer is a member of a team and carries out his or her tasks with excellence.
2. **Demonstrates Continuous Learning**. Shades of many "continuous learning" and "continuous improvement" models. There is a connection as well to Senge's learning organization [4] and overall notion of continuous improvement.
3. **Creative**. This refers to a "thinking outside the box" perspective, with special synthesis skills.
4. **Problem-Solving**. It is a great relief to have engineers who are able to actually solve real-world problems in distinction to chewing on them forever. These people are not afraid to take concrete actions in the real world, and living with the consequences.
5. **Analytic Ability**. Most engineers are strong here, but we are looking for outstanding capabilities here (not average or typical ones).
6. **Communication Skills**. Exactly, the same as an attribute from the ten characteristics list above.
7. **Logical thinking**. Highly correlated with the training and background of most engineers, especially those that have paid special attention to the mathematics portion of their studies and coursework. Possibly have had separate course in logic.
8. **Attention to Detail**. Successful managers tend to do this well, and gravitate to tasks with a lot of intrinsic detail. These are not your strategic thinkers and planners, but rather like to watch schedule and cost printouts and tend to be happiest when both are behind plans.

9. **Mathematical Ability**. Very much related to number (5) above, as well as number (7). Likes to see and tackle problem areas that could use a mathematical treatment.
10. **Leadership**. This person has many of the attributes of a leader, as articulated in this treatise as well as the rich literature. See earlier chapter on manager and leader.

NEWS PATROLLING ARTICLE [5]

This third article advertises five attributes but actually discusses more like ten desirable characteristics of a systems engineer, as below:

1. **Have a Strong Technical Foundation**. This includes, of course, all kinds of engineering courses, but also applies to physics and mathematics. At times, it reaches into chemistry, nuclear energy, and biology. Best also that it be at the master's and/or doctoral level. And do not forget the need, at times, for some background in software engineering.
2. **Be Able to Handle Multiple Responsibilities**. The systems engineer is required to be a multi-tasker in a more-or-less seamless manner.
3. **Have an Analytical Mind**. This applies to strong logical thinking which, in turn, applies to a broad systems approach.
4. **Have an Understanding of the Law**. This comes into play especially when dealing with contracts that apply to the projects that have been undertaken. Key question: how do we know that we have met the terms and conditions of the contract?
5. **Ability to Make Quality Decisions**. This can be tested and observed only after lots of decisions, and then often, it is in the eyes of the beholder. Was that a good decision, or was it not? Was it actually a decision or was it circumstances taking over?
6. **Be Able to Communicate Effectively**. This attribute appears on just about everyone's list.
7. **Must be Humble**. It's not about the engineer; it's about the project and the team.
8. **Will Need to be Self-Motivated**. This attribute ties in to having lots of curiosity that will ultimately lead to exploring a large number of issues and problems.

9. **Be Able to Diagnose Problems**. This is the first step, if you will, in problem-solving. It leads to answering the question – what is the root cause of the problem?

10. **Be Able to Implement Ideas and Listen to Customers, Clients, and Competitors**. Part of the listening feature is learning from others by being able to absorb and appropriately deal with (store and process) huge amounts of information.

REFERENCES

1. R1 – see https://specinnovations.com/10-qualities-that-make-a-good-systems-engineer/.
2. R2 – Eisner, H., "Computer-Aided Systems Engineering", Prentice-Hall, 1988.
3. R3 – McClements, D., "Ten Characteristics of Successful Engineers", see Newengineer.com, 8 January, 2019.
4. R4 – Senge, P., "The Fifth Discipline – The Art and Practice of the Learning Organization", Doubleday, 1990.
5. R5 – see www.newspatrolling.com/ "five-characteristics-of-a-systems-engineer".

Index

Printed in the United States
by Baker & Taylor Publisher Services

Printed in the United States
by Baker & Taylor Publisher Services